THE BUILDING ACTS AND REGULATIONS APPLIED

Updated 2nd Edition

THE BUILDING ACTS AND REGULATIONS APPLIED

HOUSES AND FLATS

Second Edition

C M H BARRITT

Updated to take account of
changes to Regulations and Approved Documents
since 1995

Routledge
Taylor & Francis Group

LONDON AND NEW YORK

First published 1994 by Pearson Education Limited
Second edition 1995
Updated second edition 2000

Published 2014 by Routledge
2 Park Square, Milton Park, Abingdon, Oxfordshire OX14 4RN
711 Third Avenue, New York, NY 10017, USA

First issued in hardback 2016

Routledge is an imprint of the Taylor & Francis Group, an informa business

ISBN 13: 978-1-138-15389-9 (hbk)
ISBN 13: 978-0-582-43215-4 (pbk)

British Library Cataloguing in Publication Data
A catalogue entry for this title is available from the British Library

Set by 8 in 10/12pt Baskerville No. 2

CONTENTS

PREFACE

This book has been written more from frustration than from aspiration. As a practising architect I found that I could never be sure that I had covered all the requirements of the Building Regulations as they were scattered throughout the various Approved Documents. The way the law is now set out is a considerable improvement on the way it was presented prior to 1965 and it has a logical basis in that it is all grouped according to subject. However, it is not convenient for anybody wishing to apply all the rules to a building project and I often wished that, one day, someone would write a summary that put it all together under headings related to the parts of the building.

For this reason I accepted – with some trepidation – the invitation to write a book on the Building Regulations. I am not a legal expert and the contents of this book should not be seen as an interpretation of the law. It is a re-statement of the guidance given in the Approved Documents, expanded, where appropriate, by the inclusion of British Standards recommendations where these are referred to in the Documents. I believe that this book will be of use to students struggling with the complexities of the legislation and I also hope that it will find its way on to the bookshelves of both the designers and the constructors of buildings.

I have been helped in this work by a number of people and I should like to record my appreciation of their willing assistance. The author of any textbook must have access to a comprehensive library and, in this respect, I have received valuable aid from Ian McMeekan, the site librarian at the Colchester Institute, and from the staff of the Colchester Central Library. I am grateful to them all.

I am also indebted to Colin Bassett and Basil Wilby for their very useful guidance and enthusiasm and to the staff of the Longman Group generally for the care they took in the production of this book.

My colleague and fellow architect, Brian Roach, took a very

welcome interest in what I was attempting to do and I would like to thank him for both that and the advice he gave to me.

Above all, I would like to record my appreciation of the assistance given to me by my wife, Wyn. Not only did she see that I had the time available to produce the manuscript, she also read the whole of the text looking for errors and to see whether it was easy to read and understand. For this I have the deepest gratitude – and my wife now has a wide knowledge of the Building Regulations!

PREFACE TO THE SECOND EDITION

A second edition of this book within a year of first publication has been brought about by the issue of the revised Regulations L and F and the supporting Approved Documents. These new Regulations considerably extend the application of energy conservation, introduce the need to supply an Energy Rating and amend the requirements for ventilation.

Because of the nature and form of this text, these revisions have led to changes throughout the whole book, most of which have involved rewriting portions of the text. In a lot of cases the tables and the illustrations have also been revised and a number of additional ones inserted.

At the same time, the opportunity has been taken to make a few corrections and improvements to the original text.

I am once more greatly indebted to my colleague and fellow architect Brian Roach who not only has given me much encouragement in this work but has also pointed out where a number of improvements could be made.

Updated 2ⁿᵈ edition

The 2nd edition has been updated to take account of changes to Regulations and Approved documents since 1995. Two appendices have been added:

Appendix C   Provisions for Disabled People

Appendix D   Revisions to the Regulations and Approved Documents

References to these Appendices have been inserted in the main text, so

[D1.3] Refers to section D1.3 in Appendix D, and so on.

Chapter 1

INTRODUCTION

1.1 About this book

The basis for this book is the Building Regulations 1991. These replaced the Building Regulations 1985 although not all documents published under the latter were withdrawn or changed. Some of the new or amended documents are marked '1992 Edition' and some are marked '1995 Edition' in recognition of the fact that those are the years in which the amendments came into effect, the former on 1 June 1992 and the latter on 1 July 1995.

As explained later, the ways of meeting the requirements of the Regulations are contained in fourteen 'Approved Documents', each devoted to a particular topic such as Fire Safety or Hygiene. The information given in each Approved Document relates to all types of building or building use to which the Building Regulations apply. Therefore, the requirements applicable to a single building element such as, say, a wall, are to be found distributed throughout the Documents. This book not only explains the Regulations and ways of meeting their requirements, it also has this information arranged in a more convenient form under building element headings.

Furthermore, the whole book is a selection of those aspects that relate specifically to residential accommodation.

The materials and constructional techniques illustrated in this book are intended to show the more common methods used in practice – they are not exhaustive. Other, possibly more exotic, ways of meeting the requirements of the Regulations exist and more will, no doubt, be devised in the future. The fact that they are not included in this book does not mean that they are unsuitable, merely that space did not permit them.

1.2 The development of the Building Regulations

Building control in some form has existed for a very long time. The Romans, for instance, decreed how long sun dried bricks were to be left before they could be used and also made use of a punitive system whereby if a building collapsed with fatal consequences they executed the builder!

The earliest legislation of any significance in this country was in London and followed the Great Fire of 1666. This was a building code intended, not unnaturally, to prevent the outbreak and spread of fire and enforced by 'discreet men, knowledgeable in building'. These 'discreet men' were the forerunners of the district surveyors who have, for many years, ensured compliance with the London Building Acts.

Elsewhere, there were a few towns and cities in which local Acts were passed to control fire hazards and sanitation but, generally, very little formal control existed until 1875. In that year the Government passed a Public Health Act which gave all local authorities the power to make local by-laws imposing standards of construction in relation to safety, fire prevention, health and sanitation.

Guidance was given to local authorities in the form of Model By-laws which they could adopt – with modifications as thought necessary. Housing was the only form of building subject to these rules, because this was the building type most frequently of substandard design and construction. The manner of control was to make simple statements of how the work was to be done. This assumed that by defining the way in which to build, the local authority would ensure good building performance without actually defining what that performance should be.

The introduction of building techniques using steel and concrete, which involved the mathematical analysis of structure, called for greater flexibility of control. This need gave rise to the amended Model By-laws introduced in the Public Health Act of 1936.

Following the Second World War there was a rapid development in building techniques leading to an increase in the analytical approach to building construction. There was also a much wider choice of materials and constructional systems available. As a consequence of this, many changes took place in building control, not the least significant of which was the appearance in 1953 of purely functional requirements in the by-laws. By this means the local authority could state what a system of construction must achieve rather than how it was to be built. The great advantage of this is that it left the designer free, should he so wish, to devise his own methods of building provided that he could prove that they would meet the defined standards. For the benefit of those who did not so wish, the by-laws also set out 'deemed-to-satisfy' provisions giving common building methods which, if followed, were considered capable of satisfying the mandatory performance standards.

At this time there were upwards of 1400 local authorities all with their own by-laws each containing minor but significant variations. This presented great difficulties to the many firms of architects and builders who were operating over a wide area of the country. To improve this situation, the Public Health Act of 1961 enabled the Minister to make national building regulations. The first of these came into force in February 1966 replacing all the local authority by-laws.

In 1974 there were several important Acts of Parliament, one of which was the Health and Safety at Work (etc.) Act. Its significance in this connection is that Part III of the Act took over from the Public Health Acts the legislation which enabled the appropriate Minister to make national building regulations. Not all buildings, however, were subject to these regulations. Certain buildings with specific uses were subject to their own sets of rules and, furthermore, they did not apply in London where the London Building Acts still set the standards

While these national building regulations were a great improvement on the old by-laws, they were rather difficult to understand and interpret, being written in the language reserved for legal documents as far as possible and totally without any diagrams. At the start of the 1980s both the future of building control and the form of the building regulations came under examination following which it was decided that the law, as stated in the regulations, should be separated from its interpretation as set out in the 'deemed-to-satisfy' clauses.

Following this study and ten years after the Health and Safety at Work (etc.) Act, the Building Act of 1984 took over the building control aspects of the earlier legislation and also brought together all those parts of other Acts which related to construction or design. It applied in England and Wales (including inner London from 1986) and again enabled the Secretary of State to make national building regulations and to provide documents giving guidance on them. With his new authority and the result of the research into the form of the regulations, the Secretary of State introduced the 1985 Building Regulations which consisted of a manual containing the Regulations and linking them to thirteen booklets of Approved Documents which showed ways in which the requirements could be satisfied using, for the first time, diagrams to illustrate the text.

Naturally, because of the developing nature of the industry, there were a number of amendments and revisions to the 1985 Regulations with Approved Documents being changed or withdrawn and new ones added as found necessary. Eventually, it was desirable to redraft the whole of the legislation, giving rise to the Building Regulations 1991 currently applicable. Some of the Approved Documents associated with the Building Regulations 1985 were retained in their original form but the manual was withdrawn. **D1.1** The Approved Documents that were changed are marked '1992 Edition' with the exception of two that received attention after this date and are shown as '1995 Edition'. **D1.2**

The new or revised 1992 Edition Documents are:

A   Structure
B   Fire safety
C   Site preparation and resistance to moisture
E   Resistance to passage of sound
G   Hygiene
K   Stairs, ramps and guards
M   Access and facilities for disabled people
N   Glazing – materials and workmanship
Reg 7   Materials and workmanship

The Documents revised in the 1995 Edition are:

L Conservation of fuel and power
F Ventilation

Which leaves as those that have been retained:

D Toxic substances, 1985 edition
H Drainage and waste disposal, 1990 edition
J Heat producing appliances, 1990 edition D1.3

1.3 Materials and workmanship

No matter what is promulgated in an endeavour to set desirable building standards the quality of the end result of any building exercise rests with the materials used and the workmanship employed. Regulation 7 requires that any building work is to be carried out 'with proper materials' and 'in a workmanlike manner'.

Guidance is given in the Approved Document supporting the Regulation as to how to establish the fitness of a material and, briefly, compliance with any of the following, providing that they are relevant, can be taken as indicating that they are satisfactory.

- An appropriate British Standard.
- An appropriate national standard of any Member State of the European Community or elsewhere provided that it can be shown to be equivalent to a BS.
- An Agrément Certificate issued by the British Board of Agrément.
- Any national technical approval of a Member State of the European Community orelsewhere provided that it can be shown to be equivalent to an Agrément Certificate.
- The material bears an EC mark signifying that it complies with a harmonized standard of the European Community.
- Any independent certification scheme, such as the Kitemark scheme operated by the British Standards Institution, which has been accredited by the National Accreditation Council for Certification Bodies (NACCB).
- Any tests, calculations or other means, carried out in accordance with recognized criteria, which show that the material will be capable of performing the function for which it is intended.
- Any past experience, such as a building in use, which showed that the material was sound.
- Samples taken that prove compliance with the requirements.

Short-lived materials generally are not considered suitable unless they are readily accessible for inspection and maintenance or replacement and the consequences of their failure are not likely to be serious for the health and safety of persons in and around the building.

Any material which may be harmed by the effects of moisture will meet the requirements of Regulation 7 only if the construction will keep it dry or the material is treated with some form of protection. Materials in the ground must possess the ability to resist the effects of any subsoil chemicals such as sulphates.

High alumina cement (HAC) or any product containing HAC should not be used in any structural work. Its use is only acceptable as a heat resisting material.

In certain areas of the country, the House Longhorn Beetle is liable to attack softwood in roofs. All timber must, therefore, be adequately treated to resist this infestation. The areas are:

- The Boroughs of Bracknell Forest, Elmbridge and Guildford (except the area of the former Borough of Guildford),
- The Districts of Hart (except the area of the former urban district of Fleet) and Runnymede.
- The Boroughs of Spelthorne and Surrey Heath.
- The former district of Farnborough in the Borough of Rushmoor.
- The Borough of Waverley (except the parishes of Godalming and Haslemere).
- The parishes of Old Windsor, Sunningdale and Sunninghill in the Royal Borough of Windsor.
- The Borough of Woking.

The adequacy of workmanship can be established by direct reference to a number of recognized aids such as BS 8000: *Workmanship on building sites* or equivalent technical specifications of any Member State of the European Community. It is also acceptable if it is covered by a scheme which complies with BS 5750: *Quality systems* such as one of the schemes of Registration of Assessed Capability run by the BSI.

Reference can also be made to any relevant Agrément Certificate where workmanship is specified for the material covered or any other technical approval which shows that the method of workmanship will provide an equivalent level of performance.

In this connection, probably the most frequently used criterion is past experience and this is recognized as satisfactory where, such as in a completed building, it can be shown that the method of workmanship is capable of producing a finished product which will adequately fulfil its required function.

Where the standard of workmanship is very critical, tests can be applied. One particular area where this is applicable is in drainage work and Regulation 16 allows local authorities to carry out tests as they consider necessary on sanitary pipework, drainage, cesspools, septic tanks, settlement tanks and rainwater drainage works. Guidance on how these tests should be carried out is contained in Approved Document H (see Chapter 13).

1.4 Building Regulation consent

The local authority is charged with the duty of ensuring that the law as embodied in the Building Regulations is observed. Unlike planning legislation, the process is not whether the applicant should be allowed to build but whether the proposed construction complies with the standards laid down. This makes for significant differences in the way applications are made and the way with which they are dealt.

Before proceeding to make an application or to serve notice, it is sensible to check whether such action is required. The regulations apply to the carrying out of building work in connection with any building other than those intended for certain exempt uses. The building works for which it is not necessary to obtain formal permission are listed below and shown in Figure 1.1.

- An extension of less than $30\,m^2$ to form a greenhouse, a conservatory (which must have a translucent roof), a porch, a covered way or a carport open on at least two sides (doors do not constitute an open side). If, however, the extension covers a balanced flue terminal, the Gas Safety Regulations of 1984 would apply.
- Small detached buildings of less than $30\,m^2$ floor area, built entirely of non-combustible materials, not less than 1 m from the boundary and not containing sleeping accommodation.
- A greenhouse or a building for livestock, plant or machinery.
- A movable dwelling, a tent or marquee.
- A nuclear fall-out shelter less than $30\,m^2$ in area and which is at a distance from adjoining buildings equal to the depth of the excavation plus 1 m.
- Any building not intended to remain in position for more than 28 days.
- Any ancillary buildings in connection with a building site such as the site sales office, any site offices, canteen, stores, site WCs, etc.
- The installation of heat producing gas appliances provided that it is carried out by, or under the supervision of, British Gas.

Building work consists of the erection, re-erection or extension of a building, making material alterations to a building, roofing over an open space between buildings in excess of $30\,m^2$, the conversion of movable objects (mobile homes, vehicles, vessels and the like) into a permanent building, the provision, extension or material alteration of controlled services or fittings in a building or work caused when a change of use is to take place.

The changes of use in this connection would be where a building is to be altered from some other use to a dwelling; where it will contain a flat which it did not previously have or where the change of use has removed it from one of the classes of exempt building.

1.5 The consent processes

There are two main procedures which can be followed to confirm that any proposed building work will conform to the requirements of the regulations.

The first is the traditional process of completing the necessary forms and sending them with a full set of plans to the local authority and the second is to

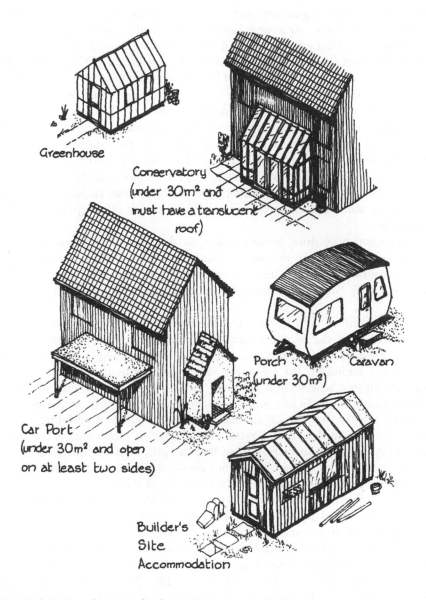

Fig. 1.1 Examples of exempt development

serve a Building Notice on the local authority informing it that work is about to be carried out.

If the first process is followed, the necessary forms can be obtained from the offices of the local authority and both they and the drawings must be submitted in duplicate plus the appropriate fee for the passing of building plans. Following their receipt, the Building Control Officer of the local authority will inspect the deposited drawings for their compliance with the standards set in the regulations and will also check any submitted calculations in connection with structural stability, thermal transmission, drainage capacities, etc. Alternatively, instead of submitting actual calculations for either of the first two, it is possible to send in a Certificate of Compliance. This must be signed by someone approved by the Secretary of State or a designated professional body. In certain specific cases the signatory must be independent and possess a prescribed indemnity insurance cover.

The local authority has five weeks in which to issue a decision. This can be extended to eight weeks with the agreement of the applicant. It is not necessary, however, to wait until the plans have been passed before work can be commenced, all that is required is to give the local authority forty-eight hours' notice. However, the Building Control Officer can require the removal of any work which is considered to be defective or not in compliance with the regulations.

The second procedure, serving notice on the local authority that building work is to be carried out, removes the burden of plan examination but not the duty of ensuring that the work is in accordance with the requirements. This is achieved by the Building Control Officer monitoring the work in progress on the site.

If this procedure is to be adopted, the necessary Building Notice form can be obtained from the local authority, completed and submitted along with a plan showing the site, the new building or building extension, other surrounding buildings, the adjoining streets and particulars of the drainage proposals all to a scale of not less than 1:1250. This must be sent in not less than forty-eight hours before any work begins on site. Clearly, as the proposals will not have the benefit of an inspection by a qualified Building Control Officer, this option does not give the same degree of protection as the first. It is essential, therefore, to make certain that the proposals not only comply fully with the latest Building Regulations but also that there are no other financial or legal encumbrances imposed on development of the site.

There is a third method, introduced in the Building Act of 1984, which is to employ an Approved Inspector. The intention of the Act was that there would be a number of inspectors, approved by the Secretary of State, in the various regions of the country who would be qualified to take over the role of the local authorities with respect to building control. As it has turned out, there is only one Approved Inspector in the country: NHBC Building Control Services Limited. This is a wholly-owned subsidiary of the National House-Building Council and it offers a complete building control service to its registered developers, builders and their designers. Legally, there appears to be no reason why this service should not be offered to all private individuals but in practice it is restricted to companies on the NHBC Register carrying out work for new or converted homes in buildings up to eight storeys high.

The procedure starts with the developer sending an Initial Notice and four copies of a site plan to his NHBC Regional Office. The NHBC Building Surveyor then completes the Initial Notice and sends it to the local authority to inform them that the NHBC will be responsible for building control on this particular project.

The local authority has ten working days in which to process and accept or reject the Notice. Once accepted, the local authority's powers to enforce the Building Regulations are suspended, the plans can be regarded as having been passed for Building Regulation and conveyancing purposes and it can also be regarded that they have been deposited for the purpose of the Advance Payments Code under the Highways Act 1980. This latter being an action which has to be taken by anybody proposing to build a road.

Following acceptance, the NHBC Building Surveyor will check the working drawings, provide advice and guidance on both the building design and the site layout and will approve the final proposals. Before building work begins, a site meeting is arranged to discuss any engineering or other problems and to enable the NHBC Inspector to familiarize himself with the builder's work programme. During the course of building operations, the Inspector will examine the construction of the foundations, the drains, the substructure and will check the completed building in response to formal notices that they are ready for such inspection. In addition he will examine other stages of the work in the normal course of his warranty inspections.

Should the Inspector discover a contravention of the regulation requirements, he will notify the developer who should arrange for the fault to be corrected within fourteen days or before the work is covered up or before the dwelling is occupied. Failure to rectify any contraventions within three months of notification can mean that the local authority's powers of enforcement are reactivated.

Unlike the local authority, NHBC Building Control is required to issue a Final Certificate when the work has been satisfactorily completed. This will form part of the Ten Year Notice under the NHBC Buildmark new home Warranty. Further details of this service can be found in the NHBC publication *Information Handbook on Building Control.*

There are a number of other organizations from whom applications for Approved Inspector status are with the Secretary of State awaiting his consideration which means that in due time, there may be more opportunity for others to take advantage of this provision in the Act.

1.6 Fees

Under the Building (Amendment of Prescribed Fees) Regulations 1990, which came into effect on 1 February 1991, a structure of fees payable to the local authority is imposed on all applicants.

If an ordinary application for approval is made, a Plan Fee must accompany the submission before the application can be considered as deposited. This Plan Fee covers the first stage of the approvals process, that of checking and passing the building plans. Following approval of the proposals, the work in progress will be inspected by a Building Control Officer to ensure that it is being carried out in accordance with both the approved plan and the Regulations. For this there is an Inspection Fee, payable immediately after the first inspection.

In the case of the second system of control, that of serving a Building Notice, the fee payable is an aggregate of both the Plan Fee and the Inspection Fee and is due as soon as work commences.

The amount of the fees is related to the size and nature of the works to be carried out and falls into one of the following three categories:

1 Extensions and alterations to domestic buildings and the erection of small garages and car ports (unless these are one of the exempt developments listed).
2 The erection of one or more dwellings.
3 Other works.

Fees in the first category, as set out in the 1994 Regulations, are fixed amounts, as shown below, if the work is of one of the types given.

Type of work	Plan fee (£)	Inspection fee (£)
An extension of less than 6 m^2	100.00	n/a
An extension between 6 and 40 m^2	60.00	150.00
An extension between 40 and 60 m^2	85.00	220.00
Loft conversion	85.00	220.00
A detached garage or car port of less than 40 m^2	17.00	53.00
Installation of cavity fill insulation not certified	n/a	50.00
Installation of unvented hot water system	n/a	50.00

Each fixed amount relates to just one item of work. If there is more than one, a corresponding fee is charged for each. For example: the Plan Fee to accompany the application for a proposal to add a 15 m^2 bedroom to a bungalow, convert the roof space into another bedroom and build a small detached garage would total £162 (£60 for the extension plus £85 for the roof conversion plus £17 for the garage. The Inspection Fee would be similarly calculated and would total £423.

Fees in this category for any work not included in the types given above and also for alterations are based on 70 per cent of a reasonable estimate of the cost.

Similarly, fees for work in the second category are calculated on the basis of a reasonable estimate of the cost, and range from £60 with no Inspection Fee for work of less than £2000, rising to a Plan Fee of £9235 and an Inspection Fee of £27 705 for work costing £10 000 000, plus £0.75 Plan Fee and £2.25 Inspection Fee, for each £1000 by which the cost exceeds £10 000 000.

The Plan Fee and Inspection Fee for work in the third category, relate to the number of dwellings, and range from £80 and £180 respectively for one house, to £600 and £2940 for 200. There are additional Plan and Inspection fees of £10 and £100 respectively for each dwelling over 200.

In all categories, value added tax is chargeable on both the Plan Fee and the Inspection Fee unless the Council's service is a statutory oblgation, in which case it is outside the scope of the legislation, and VAT is not chargeable.

NHBC Building Control Services Limited charges a fee in the same way as the local authority, based on the number of dwellings but ranging from just under £200 per dwelling where there are five houses or fewer being built simultaneously on a site to just over £100 per dwelling if the builder constructs more than 1000 per year. If a difficulty arises over contravention of the Regulations and the local authority's powers of enforcement have to be reactivated, a further fee is chargeable.

1.7 Building Notices

To make sure that the Building Control Officer is given the opportunity to carry out his inspection duties properly, the developer is required to submit Building Notices at specified stages. To facilitate this process, all Approval Notices are accompanied by a set of ten postcards which must be sent to the local authority in sufficient time to allow an inspection to be carried out. The time required is shown on each card and these are to be sent in at the following intervals:

- Commencement of the work
- Excavation for foundations
- Concrete in foundations
- Damp-proof course
- Oversite preparation
- Drains (Stage 1)
- Drains (Stage 2)
- Completion of the work
- Occupation of the building
- Installation of sanitary fittings

1.8 Relaxations

Not only are the local authorities given the powers to enforce the Regulations, they are also empowered to relax or dispense entirely with the requirements of all but a few of the Regulations should a particular requirement be shown to be unreasonable in a given circumstance. Unreasonableness is the only ground for such relaxation or dispensation.

Most of the Regulations are now of a functional nature and require provisions to be to a reasonable standard. It follows, therefore, that these cannot be relaxed since to do so would mean that the local authority was allowing something which was less than adequate and, hence, failing to apply the principle of the law. The only Regulations to which relaxations or dispensations can be applied are those with mandatory standards and it is not very likely that the application of these standards can be shown to be sufficiently unreasonable to justify a change. In consequence, this is a provision which is seldom implemented at the present time.

1.9 Energy rating

A new Regulation 14A, added in 1995, requires that any person carrying out building work that creates a new dwelling shall calculate the energy rating of the dwelling by a means of procedure approved by the Secretary of State, and shall give notice of that rating to the local authority.

The energy rating method approved is set out in Approved Document L and is expressed in terms of a score on the Government's Standard Assessment Procedure (SAP) Scale of 1 to 100, the higher the figure, the better the standard.

This SAP Rating is based on a calculated energy cost for space and water heating and will, it is claimed, inform householders of the overall efficiency of their home in a way that is both simple and easy to understand.

There is no obligation to achieve a particular SAP Energy Rating. However, in new dwellings with a Rating of 60 or less, higher levels of thermal insulation would be justified, whereas a Rating of 80 to 85 (depending on the floor area) would be taken as showing compliance with the requirements of the Regulation.

The calculation assumes a standard pattern of occupancy based on the floor area of the house and a standard heating routine. The result will depend on a range of factors that contribute to the energy efficiency of the property. These are:

• the thermal insulation of the building fabric;
• the efficiency and control of the heating system;
• the ventilation characteristics of the building;
• the solar gain characteristics of the building;
• the price of fuels used for space and water heating.

The Rating is not affected by factors that depend on the individual nature of the household occupying the premises such as:

• the household size and composition;
• the ownership and efficiency of particular electrical appliances;
• individual heating patterns and temperatures;
• the geographical location of the house.

The procedure used is based on the BRE Domestic Energy Model

(BREDEM), which provides a framework for the calculation of energy use in dwellings now updated by the 1998 edition.

Probably the best way to obtain a SAP rating is to consult a trained assessor, alternatively, one can use either a computer programme that has been developed by the Building Research Establishment, or, follow the method given in Approved Document L. Whichever is used, all applications for Building Regulation Consent in connection with a dwelling must be accompanied by either SAP rating calculations or a certificate that SAP and other calculations have been carried out correctly by a person approved by the Secretary of State for the Environment as being competent to carry out this work.

The calculation method set out in Appendix G of Approved Document L consists of a worksheet and associated tables. As the worksheet is four pages long and there are fourteen associated tables it is not an easy exercise.

A shortened form of the method is shown in Appendix A to this book and relates just to the more common forms of small new-build housing construction and servicing. If the building under consideration does not fall within the specification given in Appendix A, the full Approved Document method must be adopted.

1.10 Calculation of heat losses

Three alternative methods are given in Approved Document L to show that the heat loss through the building fabric will be limited to a level that complies with the requirements of the Regulation. They are:

- An Elemental method.
- A Target U value method.
- An Energy Rating method.

1.10.1 The Elemental method

The Elemental method consists of showing that the thermal performance of the constructional elements will meet the following standards:

	SAP rating 60 or less	SAP rating 60 or more
Roofs	0.20	0.25
Exposed walls	0.45	0.45
Exposed floors	0.35	0.45
Ground floors	0.35	0.45
Semi-exposed walls and floors	0.60	0.60
Windows, doors and rooflights	3.0	3.3

Note that any part of a roof pitched at 70° or more is counted as a wall for these purposes. For a flat roof or the sloping parts of a room in the roof a U value of 0.35 W/m²K is acceptable.

A semi-exposed wall or floor is one that separates a heated space from an unheated and uninsulated space, such as a garage.

1.10.2 The Target U value method

The Target U value method seeks to show that the average U value of the total of the area of the walls, windows, doors, etc., exposed to the outside air, plus the area of the ground floor does not exceed a Target U value derived from the following formulae:

SAP rating of 60 or less:

$$\text{Target } U\text{ value} = \frac{\text{total floor area} \times 0.57}{\text{total exposed surface area}} + 0.36$$

SAP rating of 60 or more:

$$\text{Target } U\text{ value} = \frac{\text{total floor area} \times 0.64}{\text{total exposed surface area}} + 0.4$$

The average U value is found from the formula:

$$\text{Average } U\text{ value} = \frac{\text{total rate of heat loss per degree}}{\text{total exposed surface area}}$$

The advantage of using this method is that a part of the building with a poor thermal performance can be balanced by one with a better U value (although generally the U value of exposed walls and exposed floors should not be higher than 0.7 W/m²K and that of roofs not higher than 0.35 W/m²K).

In addition, account can be taken of the heating effect of the sun through the South facing windows (solar gain) where the area of these exceeds that of the north facing windows. This is done by taking the difference between the two, calculating 40 per cent of that area and subtracting it from the total window area.

If the proposed specification does not meet the Energy Target two possible areas of change can be examined. Firstly, the specification for each of the exposed elements like the walls or roof or windows can be studied for possible improvement in the U value. Secondly, the Target U value equations assume the use of a gas or oil fired heating system with an efficiency of at least 72 per cent. If a more efficient heating system is selected the Target U value can be relaxed by up to 10 per cent.

The following calculation shows this method applied to the house shown in Fig 1.2. Note that all dimensions are to internal surfaces and the wall area excludes the windows and the area covered by the garage.

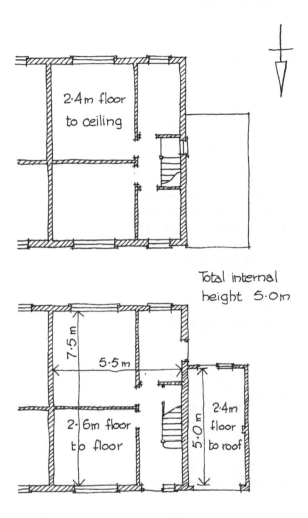

2.4m floor
to ceiling

Total internal
height 5.0m

7.5 m

5.5 m

2.6m floor
to floor

5.0 m

2.4m
floor
to roof

Fig. 1.2 Plans of a typical semi-detached house

Specification:

Walls: Brick outer leaf,
               50 mm cavity,
               600 kg/m^3 blocks,
               25 mm expanded polystyrene cavity insulation,
               13 mm lightweight plaster,
               U value $0.6 \text{ W/m}^2\text{K}$ (Table 3.3)

Roof: Timber frame,
               Tiles,
               100 mm glass fibre slab between ceiling joists
               50 mm glass fibre slab over ceiling joists
               U value $0.25 \text{ W/m}^2\text{K}$ (Table 6.18)

Floor: Hardcore,
               36 mm expanded polystyrene slabs,
               100 mm concrete,
               w.p.m.,
               50 mm screed,
               U value $0.45 \text{ W/m}^2\text{K}$ (Table 4.1)

Windows: Timber frames,
               6 mm double glazed units,
               U value $3.3 \text{ W/m}^2\text{K}$ (Table 8.2)

Doors: Solid timber panel,
               U value $3.00 \text{ W/m}^2\text{K}$ (Table 8.2)

Calculation:

	Area (m^2)	U value (W/m^{2}K)	Heat loss per degree (area × U value W/K)
Walls	66.30	0.60	39.78
Roof	41.25	0.25	10.31
Floor	41.25	0.45	18.56
Windows	14.70	3.30	48.51
Doors	3.78	3.00	11.34
	167.28		128.50

Target U *value:*

$$\frac{\text{Total floor area} \times 0.64 + 0.4}{\text{Total exposed area}}$$

$$\frac{(41.25 + 41.25) \times 0.64 + 0.4}{167.28} = 0.72 \text{ W/m}^2\text{K}$$

Average U *value:*

Total rate of heat loss per degree
Total exposed area

$\dfrac{128.50}{167.28}$ = 0.77 W/m²K

This is inadequate.

Try using 'E' glass in the double glazing units
to give a U value of 2.90 W/m²K (see Table 8.2):

Heat loss due to windows – 14.70 × 2.90 =   42.63 W/m²K

Total heat loss now   = 122.62 W/K

Average U value $\dfrac{122.62}{167.28}$ =   0.73 W/m²K

This is still inadequate.

Take account of solar gains:

Area of windows on the South	=	8.14 m²
Area of windows on the North	=	5.84 m²
Area of window on the West	=	0.72 m²
		14.70 m²

40% of 8.14 – 5.84	=	0.92
14.70 – 0.92	=	13.78
Window heat loss now 13.78 × 2.9	=	39.96
Total heat loss now	=	119.95
Average U value now $\dfrac{119.95}{167.28}$	–	0.717 W/m²K

This meets the target.

1.10.3 The Energy Rating method

As well as indicating a relative energy cost of the building, the SAP rating can be used to demonstrate that the proposals will comply with the requirements of Regulation L. All that is required is to show that the rating is not less than the appropriate figure shown below:

Floor area of house or flat (m²)	SAP Energy rating
80 or less	80
80 to 90	81
90 to 100	82
100 to 110	83
110 to 120	84
more than 120	85

Chapter 2

WORK BELOW GROUND

2.1 The Building Regulations applicable

The Building Regulations relating to work to and below ground are:

A1   Loading
A2   Ground movement
C1   Preparation of the site
C2   Dangerous and offensive substances
C3   Subsoil drainage
Reg 7 Materials and workmanship

A1 requires that the building be constructed so that it can sustain all the loads applied to it and safely transfer them to the ground; A2 requires that the building will not be impaired by any movement of the ground; C1 states that the area covered by the building must be free from vegetable matter; C2 calls for precautions against danger to health from substances present in the ground; C3 requires the provision of subsoil drainage if needed; and Regulation 7 is concerned with the use of proper materials and workmanship.

2.2 Site preparation

The Regulation requirement as shown in the Approved Document C is very simply that the area to be covered by the building is to be stripped of turf and other vegetable matter to a depth sufficient to prevent later growth. This means that the surface of the site within the area of the building must be dug away to whatever depth is necessary to remove all the topsoil capable of providing a growing medium.

2.3 Site drainage

There are two ways to deal with the prevention of moisture penetration from the ground into a building, either take the moisture away or install a barrier to its ingress. The latter is the method usually adopted and the provision of damp proof membranes is dealt with in Chapter 4. The Regulations also mention the possibility of laying subsoil drains where the water table can rise to within 0.25 m of the lowest floor of the building so that the ground water is effectively drained away by gravity. Constructing a subsoil drainage system is usually only required where the ground conditions are particularly bad and some relief of the hydrostatic pressure on the damp-proofing membranes is considered prudent.

What is probably more important is the reference in the Regulations to the action to be taken if an active subsoil drain is cut through by the building operations. Disturbance of an established flow of subsoil water can have serious and far-reaching effects. It is necessary to ensure that conditions downstream remain the same after the building is finished as they were before work started. To do this, the cut land drain pipe or pipes upstream are connected to a properly jointed pipe which is run either under or round the building and reconnected to the downstream subsoil drainage system (see Fig 2.1).

2.4 Protection against contaminants

There are very few areas of land in this country available for building purposes which have not had some previous use. Unfortunately many of these previous uses have left a legacy of contaminants which can be harmful to either the building structure or its occupants. Generally the local authority's officers will be aware of potentially contaminated sites such as landfill areas or previous industrial uses which may have allowed undesirable chemicals to soak into the ground. If this information is not available, the following site features should be checked:

Vegetation: absent or poor or unnatural growth of plants which commonly occur on vacant land may indicate the presence of metals, metal or organic compounds or gas.

Site surface: unusual colours or abrupt changes of contour may indicate deposits, wastes and residues which could contain metals, oily or tarry waste, asbestos, mineral fibres, organic compounds, combustible materials or refuse.

Fumes and odours: unpleasant smells may indicate flammable, explosive or asphyxiating gases such as methane or carbon dioxide, corrosive liquids or biologically active animal or vegetable matter.

Drums and containers: any abandoned drums on the surface could mean other containers buried in the site, leaking contaminating liquids into the ground.

If there are any signs of possible contamination, the local Environmental Health Officer should be informed immediately. If he confirms the presence of contaminants he will prescribe the action to be taken which may be removal, filling or sealing.

Removal involves taking out the contaminant and any affected ground to a

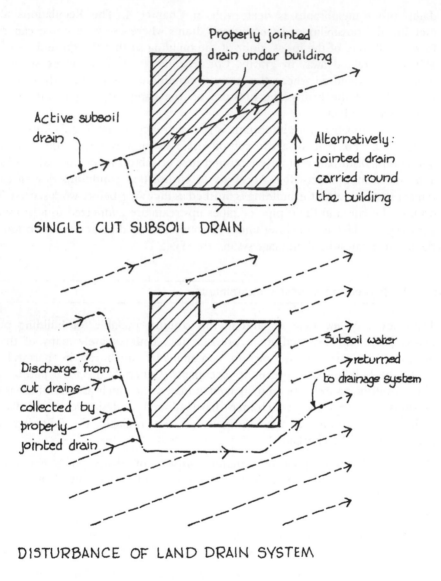

Properly jointed
drain under building

Active subsoil
drain

Alternatively:
jointed drain
carried round
the building

SINGLE CUT SUBSOIL DRAIN

Subsoil water
returned
to drainage system

Discharge from
cut drains
collected by
properly
jointed drain

DISTURBANCE OF LAND DRAIN SYSTEM

Fig. 2.1 Dealing with cut subsoil drains

depth of 1 m below the level of the lowest floor. Filling means spreading a 1 m deep layer of an inert material over the site of the building; the type of infilling and the ground floor design must be considered together. Sealing requires the installation of a suitable barrier between the building and the contamination. This must be adequately sealed at all joints, around the edges and at any service entry points. In all cases it is prudent to seek expert advice.

2.4.1 Gaseous contaminants

As well as the ground itself being contaminated, the site may be saturated with harmful gases which may be being generated away from the site. These are mostly landfill gases and, in some areas, radon.

Landfill gas arises through the breakdown of biodegradable materials in the ground and consist principally of carbon dioxide and methane. Carbon dioxide is not flammable but it is toxic; methane is an asphyxiant, it will also burn and can explode. Many of the other landfill gases are flammable and some are toxic. Methane and other gases similar to landfill gases can occur naturally and should receive the same treatment.

Radon is a naturally occurring, radioactive, colourless and odourless gas produced by the decay of uranium and radium. It can seep into buildings and increase the risk of developing lung cancer. The gas is only found in certain areas of the country which are listed in the BRE guidance document, *Radon: guidance on protective measures for new dwellings*, and are mainly Cornwall, Devon, and parts of Somerset, Northamptonshire and Derbyshire. If the presence of radon is suspected the recommendations in the BRE Report should be followed.

2.5 Foundations

Regulation A1 is summarized at the beginning of this chapter, in full it requires that the building shall be constructed so that the combined dead, imposed and wind loads are sustained and transmitted to the ground

(a) safely; and
(b) without causing such deflection or deformation of any part of the building, or such movement of the ground, as will impair the stability of any part of another building.

Regulation A2 requires that the building shall be constructed so that ground movement caused by

(a) swelling, shrinkage or freezing of the subsoil; or
(b) land slip or subsidence (other than subsidence arising from shrinkage), in so far as the risk can reasonably be foreseen,

will not impair the stability of any part of the building.

Both the design of the foundations and the nature of the subsoil on which those foundations rest must be considered to meet this requirement.

2.5.1 Stability of the subsoil

The ground below the site of the building may suffer from inherent instability due to geological faults, unstable strata, landslip, disused mines, etc., which must be taken into account when considering the nature and design of the foundations. The local authority can often advise on the possibility of such adverse conditions. In addition there are four reviews of various geotechnical conditions which have been carried out under the sponsorship of the Minerals and Land Reclamation Division of the Directorate of Planning Services of the DOE to which reference can be made. These cover research into landsliding, mining instability, natural underground cavities and foundation conditions, all in Great Britain.

2.5.2 Subsoil movement

Movement of the subsoil below a foundation only occurs if there is a change in its condition. The main constituent that can be changed is the water content and this can be reduced, increased or frozen; any of these events will threaten the stability of the building.

Reduction of the water content, in clay soils particularly, will lead to shrinkage and a general settlement of the foundations. If this is a uniform movement there may be no consequential damage to the structure (although the external underground services may suffer) but if it is greater in some places than others, cracks will inevitably develop. Great care must be taken, therefore, to avoid disturbance of any existing subsoil drainage (see Section 2.3) and the consequences of installing new land drains must be studied.

The other more common way by which subsoil water is reduced is by the action of tree roots. No direct requirements are given in the Regulations to safeguard against failure due to this cause but BS 5837 suggests that, in normal circumstances, special provisions should be made if the foundations are within a distance equal to the mature height of the tree. This, the Building Research Establishment recommends, should be increased to one and a half times the height if there is a row of trees. In the case of certain species of tree, it may be necessary to make this distance even greater. It should also be borne in mind that a large area of paving within the area of the spread of the tree will force the roots out further.

Cutting down trees in shrinkable soils does not help (even if it is permitted) because the cessation of water extraction by the roots will cause the ground to swell over a period of several years to an extent which can bring about damage to foundations not designed to accommodate ground heave.

2.5.3 Materials

Regulation 7 states that any material in the foundations must be capable of resisting attacks by deleterious substances such as sulphates. It also goes on to state that high alumina cement, which is resistant to sulphates, should not be used in structural work, including foundations.

BS 5328: Part 1, to which Approved Document A1E2 refers, makes recommendations for the types of cement, maximum free water/cement ratios and minimum cement contents which are required for different sulphate concentrations in near-neutral ground waters of pH 6 to pH 9. For very high concentrations some form of protection such as sheet polythene or a surface coating of materials such as asphalt or chlorinated rubber should be used to prevent access by the sulphate solution. The cements covered in the BS are ordinary Portland cement (OPC), OPC combined with ground granulated blastfurnace slag, OPC combined with pulverized fuel ash and sulphate resisting cement.

Where the soil is not chemically aggressive, concrete composed of normal Portland cement and fine and coarse aggregate can be used in the proportions of 50 kg of cement to not more than $0.1 \, m^3$ of sand to $0.2 \, m^3$ of coarse aggregate. Alternatively, it should be as specified in BS 5328: Part 2 for Grade ST1 concrete.

Strip foundations, as described in the next section, are usually formed in concrete, but pile foundations can, according to BS 8004, not only be either precast concrete or formed *in situ*, but can also be satisfactory in timber, particularly if it is used in the round and impregnated with a preservative.

2.5.4 Strip foundations

The loads encountered in houses and flats are usually within the bearing capacities of a shallow foundation, that is, one which transfers its load to the ground within 3 m of the surface. Depending on the type of loading, they may take the form of a pad, a raft or a strip.

Strip foundations are suitable where the loads arrive at the base of the building uniformly distributed along a line, as in the case of a load bearing wall construction. Small variations in the magnitude of the loading can be accommodated by varying the width of the strip but if there are any large differences, a different form of foundation should be considered.

There are four design considerations related to a strip foundation: the position of the wall on the strip, the width, the thickness of the concrete and its depth below ground.

Unless the foundation is specially designed, the wall must always be placed centrally on the strip to ensure that there is a uniform compressive stress on the subsoil below (see Fig. 2.2).

The width determines the amount of compressive stress on the subsoil, the wider the strip the smaller the stress. For the building to remain stable, this stress must remain within the bearing capacity of the type of subsoil while carrying the loads imposed by the structure. The precise width can be calculated from the formula:

$$\text{Minimum width} = \frac{\text{Total wall load per metre}}{\text{Allowable bearing pressure}}$$

provided that the characteristics of the ground are known. For general purposes, widths given in Table 2.1, based on Table 12 of the Building Regulations, will give a satisfactory result.

As the loads on the foundations of a bungalow are generally between 20 and 30 kN/m and on a house are between 30 and 40 kN/m, Table 2.1 has been restricted to the relevant values. For the minimum widths of strip foundations to larger buildings with loadings up to 70 kN/m see Table 12 of Approved Document A1E.

The widths given in Table 2.1 are based solely on considerations of loadbearing. It may be necessary to increase these widths to provide working space for the bricklayer building the foundation wall. In these circumstances the trench fill system can be more economical, particularly when the minimum width is less than 450 mm.

Where the thickness of the brick structure varies, such as at a chimney breast or at a pier provided to carry a particular load, the width of the foundation strip must be correspondingly varied to maintain the same projection beyond the brick face all round (see Fig. 2.2).

The depth of concrete must be such as to contain the distribution of the load of the wall out to the selected width of foundation. In this connection it is assumed that stress spreads out through concrete at an angle of 45° and therefore the depth must equal the amount of the projection beyond the wall face. To make sure that there is an adequate thickness if this projection is only small, the Regulations also impose a minimum of 150 mm (see Fig. 2.2).

The correct distance from the ground surface to the underside of the foundation is important to avoid the effects of any subsoil movement. BS 8004 states that in clay soils the strip foundations of a traditional brick building should be founded at a minimum of 900 mm below finished ground level, in practice this is generally increased to 1 m but may have to be more if there is a possibility that the subsoil may be affected by trees or hedges.

Sand and gravel types of soil are not affected by changes of subsoil water in the same way as clay. Because of this, depths of the order of 300 mm are possible. However, some of these non-cohesive soils may be subject to expansion due to frost and, for this reason, the minimum depth is usually taken as 450 mm.

As the upper surface of the strip foundation usually provides the starting point for a masonry structure it needs to be level but not all building sites are level and to avoid excessively deep excavations the strip can be stepped. If this is done, the height of the steps must not be greater than the thickness of the concrete and must be united by an overlap equal to whichever is the greatest of twice the height of the step, the thickness of the foundation or 300 mm (see Fig. 2.2).

It should be noted that, although it is not required by the Regulations nor is it mentioned in BS 8400, these step dimensions are not suitable for trench fill foundations. In this case, where the foundation thickness is 500 mm or more, it is recommended that the overlap should be the greater of either twice

Table 2.1 Minimum widths of strip foundations

Type of subsoil	Condition of subsoil	Field test	Minimum width for loads on the foundation strip of not more than		
			20 kN/m	30 kN/m	40 kN/m
Rock	Not inferior to sandstone, limestone or firm chalk	Requires at least a pneumatic pick for excavation	Equal to the thickness of the wall		
Gravel Sand	Compact Compact	Requires a pick to excavate; 50 mm square peg hard to drive beyond 150 mm	250	300	400
Clay Sandy clay	Stiff Stiff	Cannot be moulded in the fingers; removal requires a pick or pneumatic spade	250	300	400
Clay Sandy clay	Firm Firm	Can be moulded by substantial finger pressure; can be dug with a spade	300	350	450
Sand Silty sand Clayey sand	Loose Loose Loose	Can be excavated with a spade, 50 mm square peg easily driven in	400	600	Not applicable if the load per metre exceeds 30 kN
Silt Clay Sandy clay Silty clay	Soft Soft Soft Soft	Fairly easily moulded in the fingers and readily excavated	450	650	Strip widths may be calculated but expert advice is needed
Silt Clay Sandy clay Silty clay	Very soft Very soft Very soft Very soft	Natural winter sample exudes between the fingers when squeezed in the fist	600	850	

Note: for a further classification of subsoil conditions see Table 6 of BS 8004: 1986.

25

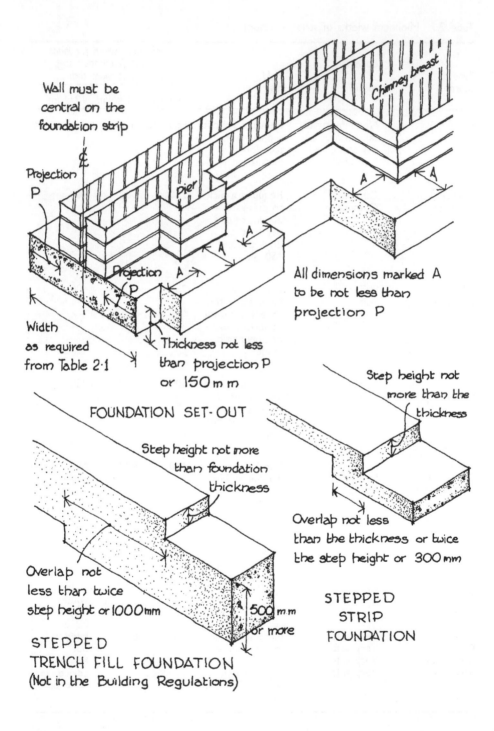

Wall must be central on the foundation strip

Chimney breast

Pier

Projection P

Projection A

P

Width as required from Table 2·1

Thickness not less than projection P or 150 mm

All dimensions marked A to be not less than projection P

FOUNDATION SET-OUT

Step height not more than foundation thickness

Step height not more than the thickness

Overlap not less than the thickness or twice the step height or 300 mm

Overlap not less than twice step height or 1000 mm

500 mm or more

STEPPED TRENCH FILL FOUNDATION
(Not in the Building Regulations)

STEPPED STRIP FOUNDATION

Fig. 2.2 Foundation strips

the step height or 1 m and the step height should never be more than the foundation thickness.

2.5.5 Foundations to framed structures

Unlike a brick building, the loads from a framed structure, such as may be found in a multistorey block of flats, arrives at the foundation level in concentrated points. To transfer these concentrated loads to the ground safely, as required in the Regulations, pad foundations can be used if the ground conditions are suitable.

The design principles are similar to strip foundations in that the size of the pad is derived from the load to be carried and the bearing capacity of the ground and should be carefully calculated. Invariably the pads are square in plan even though the load distributes through the concrete in a cone shape, mainly because a square shape is easier to dig.

The thickness of unreinforced pad foundations must contain the 45° spread of load from the base of the structural member. If the pad is reinforced, its thickness and reinforcement must be designed to resist the bending forces applied and provide adequate cover to the steel.

2.5.6 Raft foundations

Where the ground bearing capacity is low or there is a risk of differential settlement the consideration of the use of a reinforced concrete raft is recommended, particularly for lightly loaded structures such as low rise dwellings. It can also happen that the pads under the point loads of a framed structure on poor ground occupy so much of the available area that it makes practical sense to join them up into a raft.

The analysis of the forces involved and the provisions to be made to meet them in a raft foundation is not a simple matter and, consequently, each one must be individually designed for the specific set of circumstances.

2.5.7 Pile foundations

BS 8004 recommends that the use of short piles should be considered where the ground conditions are poor or unstable, particularly in connection with dwellings and lightly loaded framed structures. The conditions making the method a viable alternative are very soft or made ground and clay soils where swelling or shrinkage can be anticipated.

The type, method of construction, size and load capacity must be carefully considered and it is recommended that specialists should be consulted for guidance on the many types of technique available.

Chapter 3

EXTERNAL WALLS

3.1 The Building Regulations applicable

The Building Regulations relating to the construction of the external walls of a residential building are:

A1   Loading
B4   External fire spread
C4   Resistance to weather and ground moisture
D1   Cavity insulation
L1   Conservation of fuel and power
Reg 7 Materials and workmanship

A1 requires that the building, in this case the walls, will be constructed so that the imposed loads are sustained safely; B4 states that the external walls of the building must resist the spread of fire from one building to another; C4 deals with provisions designed to keep the interior of the building dry; D1 is concerned that reasonable precautions are taken to prevent the permeation of toxic fumes from insulation in a cavity into the building; L1 very simply requires that reasonable provision be made to conserve fuel and power in a building; and Regulation 7 is about materials and workmanship.

In addition, a set of Building Regulations includes a copy of the BRE Report *Thermal insulation: avoiding risks*. This is not an Approved Document but its recommendations are given where appropriate.

3.2 Structural stability

To be stable, the external walls of a non-framed building must sustain the dead loads of the rest of the structure plus the live loadings due, firstly, to the occupancy of the building and, secondly, to wind pressure. Section 1C of Regulation A1 gives values for the thickness of the walls of residential buildings up to three storeys which lie within the proportions shown in Fig. 3.1 and comply with the conditions given. If these are followed the wall should

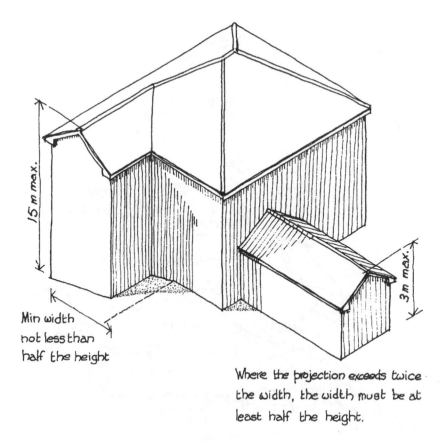

15 m max.

Min width
not less than
half the height

3 m max.

Where the projection exceeds twice
the width, the width must be at
least half the height.

Fig. 3.1 Proportions of residential buildings

be capable of withstanding the worst combination of circumstances likely to arise in this country.

3.2.1 Conditions

The wall thicknesses given in Table 3.2 are based on the following assumptions:

1 The floor enclosed by structural walls does not exceed 70 m².
2 The imposed loads do not exceed:
   (a) roofs up to 6 m span: 1.5 kN/m²;
   (b) roofs from 6 to 12 m span: 2.0 kN/m²;
   (c) floors: 2.0 kN/m² – span not to exceed 6 m between centres of bearings;
   (d) ceilings: 0.25 kN/m² plus 0.9 kN concentrated load.
3 The design wind speed is no greater than 44 m/s. In England and Wales this speed may be exceeded in the North, West and South West as shown on the map in Fig. 3.2.

In these areas, depending on the location of the site and whether it is flat or

Gust speeds likely to be exceeded only once in 50 years at 10m above open level country

Fig. 3.2 Map of basic wind speeds

steeply sloping, the height of the building may need to be restricted. On level or slightly sloping sites in open countryside where there are windbreaks and where the basic wind speed is 48 m/sec, the maximum height is limited to 13 m. Without the windbreaks, the heights would be restricted to 11 m where there is a basic wind speed of 46 m/sec and 9 m where the speed is 48 m/s. The maximum height on steeply sloping sites and cliffs is further restricted in most locations except sheltered city centres as shown in Table 3.1.

3.2.2 Wall thicknesses

The basic requirement is that the minimum thickness of a solid wall in coursed brickwork or blockwork is one-sixteenth of the storey height. This

Table 3.1 Maximum height of buildings on steeply sloping sites and cliffs

Basic wind speed (m/s)	Maximum building height in metres for sites located in:			
	Open countryside with no windbreaks	Open countryside with some windbreaks	Countryside with many windbreaks. Small towns City suburbs	Protected city centres
36	8	11	15	15
38	6	9	15	15
40	4	7.5	14	15
42	3	6	12	15
44	n/a	5	10	15
46	n/a	4	8	15
48	n/a	3	6.5	14

Note: on unprotected sites in open countryside with wind speeds in excess of 44 m/s the Building Regulation guidance does not apply.

(*Source:* Based on Table 9 of Approved Document A1)

basic requirement is modified in relation to the height and length of the wall, as shown in Table 3.2, with the effect that the basic standard rarely applies. This table does not apply to any walls built as part of a bay window above the level of the ground floor sill.

In calculating the minimum thicknesses given for the leaves of a cavity wall the following brick and block compressive strengths are assumed:

1 One or two storey building
   (a) each leaf: bricks of $5 \, \text{N/mm}^2$ or blocks of $2.8 \, \text{N/mm}^2$.
2 Three storey building
   (a) outer leaf: bricks or blocks of $7 \, \text{N/mm}^2$.
   (b) inner leaf, lowest storey: bricks of $15 \, \text{N/mm}^2$ or blocks of $7 \, \text{N/mm}^2$.
   (c) inner leaf, upper storeys: bricks of $5 \, \text{N/mm}^2$ or blocks of $2.8 \, \text{N/mm}^2$.

Furthermore, the mortar should be to the proportions either as given in BS 5628: Part 1 for mortar designation (iii) or 1:1:6 Portland cement, lime and sand measured by volume of dry materials.

If it is a solid wall of uncoursed stone, flints, clunches or similar less stable materials, the minimum thickness must be 1.33 times the thickness given in Table 3.2.

The cavity in a cavity wall must be at least 50 mm wide with wall ties at 900 mm spacing horizontally and 450 mm spacing vertically. If the cavity width is increased to between 75 and 100 mm, the horizontal wall tie spacing must be decreased to 750 mm and vertical twist ties should be used

Table 3.2 Minimum thickness of walls built of coursed brickwork or blockwork

Height of wall	Length of wall	Storey	Minimum thickness of a solid wall	Minimum thickness of the leaves of a cavity wall
Not exceeding 3.5 m	Not exceeding 12 m	All storeys	190 mm	90 mm each leaf
Exceeding 3.5 m but not exceeding 9 m	Not exceeding 9 m	All storeys	190 mm	90 mm each leaf
	Exceeding 9 m but not exceeding 12 m	Lowest storey*	290 mm	280 mm total but neither leaf to be less than 90 mm
		All upper storeys	190 mm	Each leaf 90 mm
Exceeding 9 m but not exceeding 12 m	Not exceeding 9 m	Lowest storey*	290 mm	280 mm total but neither leaf to be less than 90 mm
		All upper storeys	190 mm	90 mm each leaf
	Exceeding 9 m but not exceeding 12 m	Lowest two storeys*	290 mm	280 mm total but neither leaf to be less than 90 mm
		All upper storeys	190 mm	90 mm each leaf

(*Source:* Based on Table 5 of Approved Document A1 and paragraph IC8)

* The height of the lowest storey is measured from the base of the wall.

conforming to BS 1243. It should be noted that if the cavity is to be partially filled with thermal insulation, the space left (the 'residual cavity') must be not less than 50 mm.

The length of the wall is measured between the centres of the buttressing walls, piers or chimneys which provide end restraint (see Section 3.2.4). The height is measured from the top of the foundations to the eaves or half way up the gable, whichever is the case.

3.2.3 Differences in ground level

The thickness of the wall must be adjusted so that the height of any difference between the level of the ground each side of a wall or between the ground level and the floor level is not more than four times the thickness of the wall – if it is a solid wall, or four times the sum of the thicknesses of the leaves in a cavity wall without any cavity filling. In this respect it is also necessary to consider the position of the damp proof course as defined in Approved Document C4 (see Section 3.3 below)

3.2.4 End restraint

The length of a wall for the purposes of Table 3.2 is measured between centres of supports. These supports can be in the form of piers, buttressing walls or chimneys provided that they meet the criteria set out below and in Fig. 3.3.

It is assumed that the buttressing walls covered in the details shown in Fig. 3.3 are themselves supported by other walls, piers or chimneys. If they are not, then their thickness should be not less than:

1 half the thickness required in Table 3.2 for a wall of similar height and length, less 5 mm;
2 75 mm if the wall is less than 6 m high and 10 m long;
3 90 mm in any other case

3.2.5 Lateral support

External walls should be provided with lateral support at each storey level and at the top. This can be provided in houses of not more than two storeys:

1 At the first floor level by:
    (a) floor joists bearing on to the wall by at least 90 mm at centres not less than 1.2 m;
    (b) floor joists at a maximum of 1.2 m centres bearing at least 75 mm on to a wall plate built into the wall;
    (c) joists bearing on to hangers of the restraint type described in BS 5628: Part 1 which are hooked through the wall at not less than 2 m centres;
    (d) a concrete floor bearing at least 90 mm into the wall.
2 At the eaves level by the framing of the roof if it is pitched at 15° or more, is tiled or slated, has main timber members bearing on to the wall at not less than 1.2 m centres and is of a type known to be resistant to gusts of wind.

Walls running parallel to the floor joists, gable walls and walls of buildings with flat or low pitched roofs should be strapped to the floor or roof as shown in Fig. 3.4 with galvanized mild steel straps of a minimum cross-section of 30 × 5 mm fixed across three joists or rafters at not less than 2 m centres. The joist or rafter adjacent to the wall should be solidly blocked against the wall

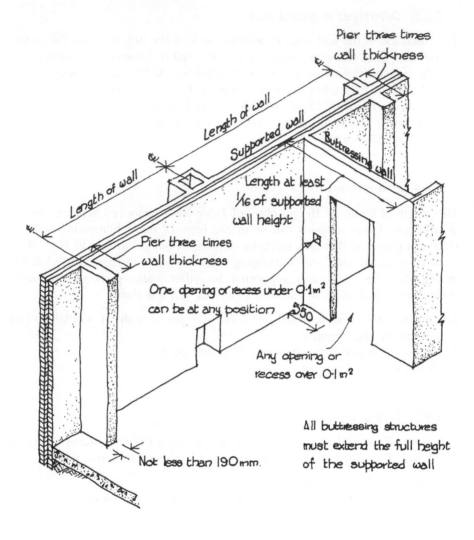

Pier three times
wall thickness

Length of wall

Supported wall

Length of wall

Buttressing wall

Length at least
1/16 of supported
wall height

Pier three times
wall thickness

One opening or recess under 0·1 m²
can be at any position

Any opening or
recess over 0·1 m²

All buttressing structures
must extend the full height
of the supported wall

Not less than 190 mm.

See figure 3·6 for openings in supported walls

Fig. 3.3 Buttressing structures

face. Noggins, equal to at least half the depth of the joist and 38 mm thick, should be fixed between the strapped joists or rafters.

Where an opening, such as a stairway, is provided in a floor or roof and the lateral support which would otherwise have been provided is interrupted, the length of the opening, measured parallel to the wall, should be limited to 3 m and the spacing of any restraint anchors reduced so that the same number is used as would be provided if the opening had not been there. If the restraint is provided by joists built into the wall or roof timbers bearing on to the wall,

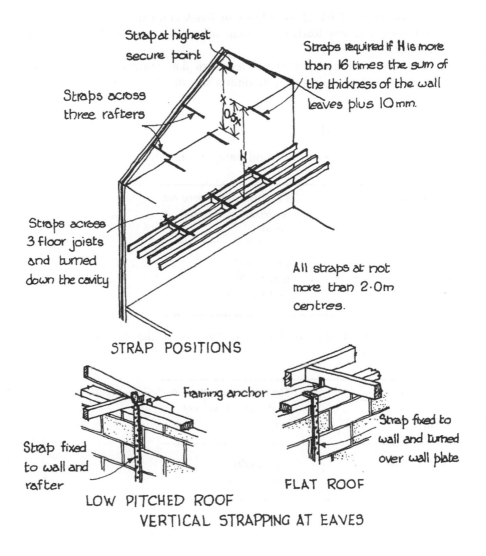

Strap at highest
secure point

Straps required if H is more
than 16 times the sum of
the thickness of the wall
leaves plus 10 mm.

Straps across
three rafters

0.5x

X

H

Straps across
3 floor joists
and turned
down the cavity

All straps at not
more than 2·0m
centres.

STRAP POSITIONS

Framing anchor

Strap fixed
to wall and
rafter

LOW PITCHED ROOF

Strap fixed to
wall and turned
over wall plate

FLAT ROOF

VERTICAL STRAPPING AT EAVES

Fig. 3.4 Lateral support by straps

this should be continuous for the whole length of the wall on each side of the
opening.

3.2.6 Single leaf external walls

Garages and similar single storey buildings not exceeding 3 m high or 9 m
wide, or annexes not more than 3 m high, neither of which is for residential
purposes or encloses a floor area greater than $36 \, m^2$, can have single leaf
external walls which should be:

- at least 90 mm thick of solid brick or block construction;
- not subject to any loads other than wind and those of the building's own roof;
- supported each end and at not less than 3 m centres by a buttressing wall at right angles or piers of a minimum size of 190 × 190 mm.

3.2.7 Parapet walls

The minimum thicknesses for solid parapet walls are:

Thickness (mm)	Parapet height above the underside of the roof (mm)
150	up to 600
190	600 to 760
215	760 to 860

The minimum for the combined thicknesses of the two leaves of a cavity parapet wall are:

Thickness (mm)	Parapet height above the underside of the roof (mm)
under 200	up to 600
200 to 250	600 to 860

3.3 Damp-proofing

The Regulations require that an external wall must not only resist the passage of moisture from the ground and the penetration of rain and snow, it must also not be damaged by ground moisture or rain or snow. The latter is a matter of choice of materials, rising ground moisture is stopped by a damp-proof course and horizontal penetration by either the materials or the construction.

3.3.1 Damp-proof courses

The products mentioned in Approved Document C4 as suitable for damp-proof courses are: bituminous material, engineering bricks and slates in cement mortar but any other material that will prevent the paasage of moisture would be acceptable.

The damp-proof level must be not less than 150 mm above the adjoining ground and continuous with the floor damp-proof membrane. If it is a cavity wall the damp-proof coursing in each leaf does not have to be at the same level but the cavity must be carried down at least 150 mm below the lowest d.p.c. or, alternatively, it can be arranged as a damp-proof tray fixed so as to direct any water in the cavity back to the external face.

Walls subject to ground water pressure such as basement walls and situations where the Regulation standards are not appropriate should have damp-proofing measures as recommended in Clauses 4 and 5 of BS 8215: *Code of Practice for design and installation of damp-proof courses in masonry construction* or BS 8102: *Code of practice for protection of structures against water from the ground.*

Damp-proof courses should also be provided where the downwards flow of moisture would be interrupted by an obstruction such as a lintel and below an opening unless there is a sill which will provide a complete barrier. The damp-proof course in this case should be arranged to direct the water to the outside face.

3.3.2 Rain penetration in solid walls

A solid wall can be acceptable in its ability to prevent the penetration of rain to the interior of the building provided that it is thick enough to hold the moisture until it can be released to the outside in a dry period before it either reaches the inside face or causes damage to the structure. This is dependant on the type of walling used and the severity of exposure of the wall.

The Approved Document recommends reference in this case to the BSI Draft for Development DD 93: *Method for assessing exposure to wind-driven rain* and BS 5628: *Code of practice for use of masonry*, Part 3: *Materials and components, design and workmanship*. Also mentioned is the BRE publication *Thermal insulation – avoiding the risks*. This latter publication is now supplied as part of the package of Approved Documents, although it is not itself an Approved Document.

As guidance, the Approved Document states that in conditions of severe exposure a solid wall may be of brickwork at least 328 mm thick, dense aggregate blockwork at least 250 mm thick or lightweight blockwork at least 215 mm thick, any of which should have the joints raked out to a depth of 10 mm and be rendered.

The rendering is to be in two coats to a total thickness of not less than 20 mm with a scraped or textured finish. Its strength should be compatible with the strength of the walling material but, generally, should be: 1 part cement to 1 part lime to 6 parts well graded sharp sand unless the wall is of dense concrete blocks in which case the mix should be $1 : \frac{1}{2} : 4$. Where the severity of exposure or a different type of block requires it, the Document recommends BS 5262: *Code of practice. External rendered finishes.*

Approved Document C4 also refers to BS 5390: *Code of practice for stone masonry* where this is applicable.

3.3.3 Cavity walls

Any cavity wall will meet the requirements if it is built with an outer leaf of bricks, blocks, natural stone or cast stone and an inner leaf of similar masonry or of framing with a lining. It should have a cavity at least 50 mm wide which should be maintained even when it is partially filled with thermal insulating material.

3.4 Thermal insulation

The insulation value of the building fabric is measured by its U value. This is its capacity to transfer heat from the inside to the outside face. It is expressed as W/m^2K (watts per square metre per degree kelvin). (Kelvin is used for the difference between one given temperature and another whereas Celsius is used for the difference between zero and a given temperature.) A structure with a U value of, say, $1\,W/m^2K$ will transfer one watt of heat energy through each square metre of its area for every degree of temperature difference between the faces.

The Building Regulation L1 states that 'reasonable provision shall be made for the conservation of fuel and power in buildings' and Section 1 of Approved Document L1 states that this requirement will be met in relation to wall structures if they have a maximum U value of 0.45 W/m^2K. These values apply whatever the SAP rating of the building.

This, plus other standards related to the floor and roof (see Chapters 1, 4 and 6) is the 'elemental approach'.

3.4.1 Wall constructions

The Approved Document gives a guide to working out the thickness of insulating material required to achieve the U value of 0.45 W/m^2K required in external walls (and the 0.6 W/m^2K required for semi-exposed walls). The guide is contained in Tables A5, A6, A7 and A8. Table 3.3 is an extraction of these four tables covering the more common materials and constructions. Interpolation may be used for concrete blocks with a density between the values given and extrapolation may be used for blocks exceeding 100 mm thick.

To use Table 3.3, first select the insulation to be used either by its description or by its thermal conductivity, then read off the base thickness

Table 3.3 Thicknesses of insulation (mm) for a design *U* values of 0.45 W/m²K and 0.6 W/m²K

Insulation material		Phenolic foam		Polyurethane board		Expanded polystyrene slab, Glass fibre slab, or Mineral fibre slab		Glass fibre quilt, or Urea formaldehyde foam	
Thermal conductivity	(W/mK)	0.020		0.025		0.035		0.040	
Design U value of wall	W/m² K	0.45	0.6	0.45	0.6	0.45	0.6	0.45	0.6
Base thickness of insulation	mm	41	30	51	37	71	52	82	59

Allowable reductions for wall components:

Brick outer leaf		2	3	4	5
100 mm Concrete outer leaf:	kg/m³				
Aerated	600	8	10	15	17
	800	7	8	12	14
Lightweight	1000	6	7	10	11
Dense	1600	3	3	5	5
100 mm Concrete inner leaf:	kg/m³				
Aereated	600	9	11	15	17
	800	7	9	13	15
Lightweight	1000	6	8	11	12
Dense	1600	3	4	5	6
50 × 100 nom. Timber frame and glass fibre slab		42	53	74	84
50 × 100 nom. Timber frame and glass fibre quilt		38	48	67	77
Cavity (25 mm minimum)		4	5	6	7
13 mm plaster		1	1	1	1
13 mm lightweight plaster		2	2	3	3
10 mm plasterboard		1	2	2	3
13 mm plasterboard		2	2	3	3
Airspace behind plasterboard dry lining		2	3	4	4
9 mm sheathing ply		1	2	2	3
20 mm cement render		1	1	1	2
13 mm tile hanging		0	0	1	1

(*Source:* Tables A5, A6, A7 and A8 of Appendix A to Approved Document L: 1995 Edition)

below the appropriate design *U* value. Continue down the column to find the allowable reductions for the wall components and subtract these from the base thickness. The result is the minimum thickness of insulation required. The example below shows the calculation for the filled cavity wall construction shown in Fig. 3.5:

Example

Insulation material: glass fibre quilt		
Base thickness for *U* value of 0.45W/m²K	=	82.0 mm

Allowable reductions:

Brick outer leaf	=	5	mm
Block inner leaf 700 kg/m³ (interpolated between 17 and 15)	=	16	mm
Plasterboard	=	3	mm
Total deduction from base thickness	=	24	mm

Minimum thickness of insulation required: 82 – 24	=	**58 mm**

Nearest available thickness 60 mm

There are innumerable ways of building external walls to satisfy the requirements of this legislation, particularly with the use of cavity walls where any necessary insulants can be added to the external face, the internal face or, as is commonly done, placed in the cavity. Most manufacturers of walling materials, particularly the lightweight concrete block industry, produce guides of how to use their products so that the wall meets the standards. Figure 3.5 shows twelve examples of solid walls and cavity walls with unfilled, partially filled and filled cavities all of which will achieve the insulation standard indicated.

Thermal insulating material may be inserted in the cavity of an external wall under the following conditions:

1 Any rigid material should be the subject of a current certificate of the British Board of Agrément and its installation carried out in accordance with the requirements of that document.
2 Urea-formaldehyde foam can be inserted into a wall after it has been built. The inner leaf must be of bricks or blocks and the work should be in accordance with BS 5617: *Specification for urea-formaldehyde (UF) foam systems suitable for thermal insulation of cavity walls with masonry or concrete inner and outer leaves.* It should be installed in accordance with BS 5618: *Code of practice for thermal insulation of cavity walls by filling with urea-formaldehyde foam systems.* The suitability of the wall for foam filling should be assessed before the work is carried out and the work must be done by a holder of a current BSI Certificate of Registration of Assessed Capability or a similar document issued by an equivalent body.
3 Other insulating material inserted into the cavity after the wall has been built should be installed in accordance with BS 6232: *Thermal insulation of cavity walls by filling with blown man-made mineral fibre,* Part 1: 1982: *Specification for the performance of the installation systems* and Part 2: 1982: *Code of practice for installation of blown man-made mineral fibre in cavity walls with masonry and/or concrete leaves.* The suitability of the wall before filling should be assessed and the person doing the work should hold a current BSI Certificate of

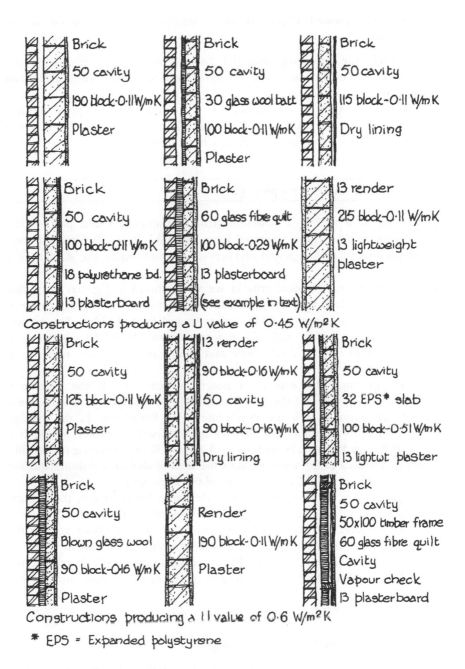

Brick

50 cavity

190 block-0·11 W/mK

Plaster

Brick

50 cavity

30 glass wool batt

100 block-0·11 W/mK

Plaster

Brick

50 cavity

115 block-0·11 W/mK

Dry lining

Brick

50 cavity

100 block-0·11 W/mK

18 polyurethane bd.

13 plasterboard

Brick

60 glass fibre quilt

100 block-0·29 W/mK

13 plasterboard

(see example in text)

13 render

215 block-0·11 W/mK

13 lightweight plaster

Constructions producing a U value of 0·45 W/m²K

Brick

50 cavity

125 block-0·11 W/mK

Plaster

13 render

90 block-0·16 W/mK

50 cavity

90 block-0·16 W/mK

Dry lining

Brick

50 cavity

32 EPS* slab

100 block-0·51 W/mK

13 lightwt plaster

Brick

50 cavity

Blown glass wool

90 block-0·16 W/mK

Plaster

Render

190 block-0·11 W/mK

Plaster

Brick

50 cavity

50x100 timber frame

60 glass fibre quilt

Cavity

Vapour check

13 plasterboard

Constructions producing a U value of 0·6 W/m²K

* EPS = Expanded polystyrene

Fig. 3.5 Thermal insulation of walls

41

Registration of Assessed Capability or a similar document issued by an equivalent body.

Alternatively the insulating material and its installation should be the subject of a current British Board of Agreement certificate or a European Technical Approval.

If the construction is of timber, it will incorporate vapour control membranes and, since leakage of cold outside air significantly affects the heating demand, these membranes should be sealed.

3.5 Window and door openings

Any opening in a wall is a potential structural weakness and part of the interpretation of the requirements of Regulation A1 states that the number, size and position of openings and recesses should not impair the stability of a wall.

To achieve this, dimensional criteria are set as shown in Fig. 3.6 and Table 3.4. It should be noted that the total width of openings and recesses between two points of lateral support of the wall must not exceed either 3 m or two-thirds of the length of the wall. These requirements should be compared to those described in Section 3.2.4 above and shown in Fig. 3.3 for opening in buttressing walls.

Window openings also present a potential heat loss. If the Elemental Method is used to show that the heat loss through the building fabric of a new house or flat complies with the requirements of Regulation L, the average U value of the windows, doors and any rooflights should not exceed 3.0 W/m^{2}K (if the SAP Energy Rating is 60 or less) or 3.3 W/m^{2}K (if the Rating is over 60). In addition, the area of the windows, plus the area of all external doors and any rooflights should not exceed 22.5 per cent of the floor area.

If the proposal is for an extension to an existing house or flat, the average U value of the windows, doors and rooflights should not exceed 3.3 W/m^{2}K and the area should not exceed 22.5 per cent of either, the floor area of the extension itself, or, the floor area of the dwelling plus the extension.

See Chapter 1: Sections 1.9 and 1.10 for explanations of SAP Energy Rating and the Elemental Method and Chapter 8 for types of windows and their U values.

3.5.1 Thermal bridging around openings

The cavity must be closed around all window and door openings and in doing so, a potential path of heat loss is created and a consequential problem of local condensation occurring. Approved Document L states that provision must be made to limit the amount of heat lost via this thermal bridging.

Figure 3.7 shows the constructional methods recommended in the

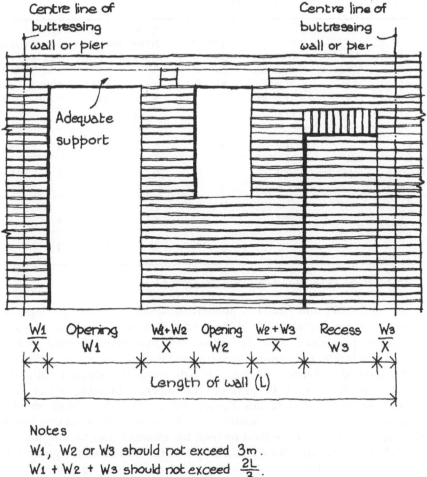

Centre line of buttressing wall or pier

Centre line of buttressing wall or pier

Adequate support

| $\dfrac{W_1}{X}$ | Opening W_1 | $\dfrac{W_1 + W_2}{X}$ | Opening W_2 | $\dfrac{W_2 + W_3}{X}$ | Recess W_3 | $\dfrac{W_3}{X}$ |

Length of wall (L)

Notes

W_1, W_2 or W_3 should not exceed 3m.

$W_1 + W_2 + W_3$ should not exceed $\dfrac{2L}{3}$.

See table 3·4 for the values of factor X.

Wall lengths between openings or lines of support are minimum dimensions.

See figure 3·3 for positions of openings in buttressing walls.

Fig. 3.6 Dimensional criteria for openings and recesses

Approved Document that, if followed will satisfy this requirement. It should be noted that, in all cases, the details given involve setting the frame further back in the wall than has been the practice in recent years and make it necessary to fit a sub-cill.

An alternative to adopting one of these constructions is to show, by calculation, that the edge details will give a satisfactory performance. This can be done either by following the procedure given in BRE IP 12/94 *Assessing*

Table 3.4 Value of factor 'X' in Fig. 3.6

Nature of roof span	Maximum span (m)	Minimum thickness of wall inner leaf (m)	Span of floor is parallel to wall	Span of timber floor into wall		Span of concrete floor into wall	
				max 4.5 m	max 6.0 m	max 4.5 m	max 6.0 m
				Value of factor 'X'			
Roof spans parallel to wall	not applicable	100	6	6	6	6	6
		90	6	6	6	6	5
Timber roof spans into wall	9	100	6	6	5	4	3
		90	6	4	4	3	3

(*Source:* Based on Table 10 of Approved Document A1/2)

condensation risks and heat loss at thermal bridges around openings, or by adopting the procedure given in Appendix D to the Approved Document. This latter procedure aims to assess the minimum thermal resistance between the inside and outside surfaces at the edges of openings by adding up the thermal resistances of the materials along the shortest path. If this resistance is less than 0.45 m^{2}K/W, the thermal bridging effect can be ignored but if it exceeds this figure, either the edge detail should be changed to improve the resistance or compensating measures, such as reducing the U value of the walls or roof, for example, can be introduced.

Metal lintels present a particular case in that the thin metal is a good conductor and increases the risk of condensation and mould growth. Additional calculations are needed to check this, as shown in the Approved Document.

3.5.2 Chases

In addition to openings, reduction of the wall thickness by chases can impair its stability. These should be carefully positioned, especially in hollow block walls. Vertical chases should not be deeper than one-third of the wall thickness or, in cavity walls, one-third of the leaf thickness and horizontal chases should be less than one-sixth of the thickness of the wall or leaf.

3.6 Fire safety

The restriction of the spread of fire from one building to another was the subject of the earliest legislation on building control (see Chapter 1) and, in

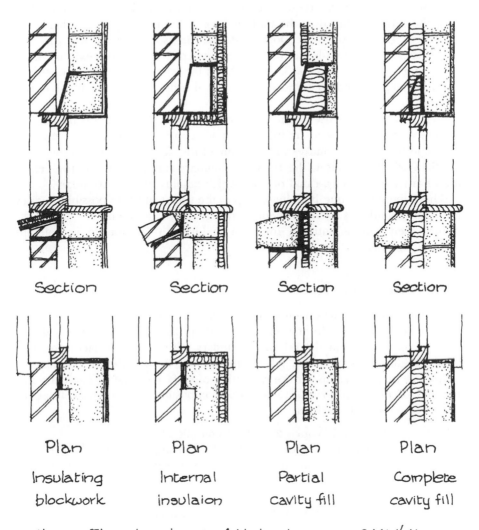

Section   Section   Section   Section

Plan   Plan   Plan   Plan

Insulating   Internal   Partial   Complete
blockwork   insulaion   cavity fill   cavity fill

Notes: Thermal conductivity of blockwork not over 0·16 W/mK
      Frame overlap onto blockwork:
          30 mm for dry-lining
          55 mm for lightweight plaster
      Steel lintels finished with min 15 mm
      lightweight plaster or dry-lined.

Fig. 3.7 Constructions to limit thermal bridging at openings

principle, nothing much has changed except the way in which the law is expressed. Regulation B4 requires that 'The external walls of a building shall resist the spread of fire over the walls and from one building to another . . .'

Section 12 of Approved Document B4 makes provisions for the fire resistance of external walls and to limit the susceptibility of their external faces to ignition and flame spread. Section 13 sets limits on the extent of openings and other unprotected areas in external walls in order to reduce the risk of fire spread by radiation.

Both sections make use of the term 'relevant boundary'. This is usually the boundary of the site faced by the wall under consideration. To be 'relevant' any angle the boundary makes with the line of the side of the building must be less than 80°. In addition, the relevant boundary can be taken as the centre line of any road, railway, canal or river which adjoins the site boundary.

Where more than one building occupies the site, it must be possible to draw 'notional' boundaries between them such that the requirements for all aspects of fire safety can be satisfied for each building in relation to its relevant notional boundary.

3.6.1 Wall construction

Two aspects of construction must be considered: surface spread of flame and inherent fire resistance. In any building less than 20 m high, any external wall face less than 1 m from the relevant boundary must have a surface spread of flame classification of Class O. Where the distance to the relevant boundary is more than 1 m, provisions for the restriction of surface spread of flame are not required. Class O materials are any non-combustible materials or those of limited combustibility such as:

- brickwork, blockwork, concrete and ceramic tiles;
- plasterboard (painted or not, or with a PVC facing not more than 0.5 mm thick), with or without an air gap or fibrous or cellular insulating material behind;
- woodwool cement slabs;
- mineral fibre tiles or sheets with cement or resin binding.

Inherent fire resistance may fail in one of three ways: the load bearing capacity of the wall is lost and the structure collapses; the fire breaks through the wall (loss of 'integrity'); or the ability of the wall to give insulation from the effects of the heat of the fire is destroyed. Since almost nothing is fire proof the Regulations lay down the length of time of resistance appropriate for specific building uses in given situations.

The following are minimum periods of fire resistance of the load bearing capacity, integrity and insulation of walls less than 1 m from the boundary exposed to fire from either outside or inside:

- Houses
  - up to 5 m high   30 minutes
  - 5 to 20 m high   60 minutes
- Flats and maisonettes
  - up to 5 m high   30 minutes
  - 5 to 20 m high   60 minutes
  - 20 to 30 m high   90 minutes
  - over 30 m high   120 minutes
- Flat conversions (with adequate means of escape)
  - up to three storeys   30 minutes
  - four storeys and over   as for flats above
- Basements in
  houses, not more than 10 m deep   30 minutes
  - flats, not more than 10 m deep   60 minutes

Walls over 1 m from the boundary are required to resist the effects of fire only from the inside of the building. The resistance periods of their load bearing capacity and integrity are to be as shown above but their insulation resistance time is reduced to 15 minutes.

In tall buildings, i.e. with a storey at 15 m above ground level, any thermal insulating material used in an external wall should be of limited combustibility unless it is in a cavity wall of masonry construction where the insulation totally fills the cavity or, if it partially fills it, the cavity is closed at the top.

3.6.2 Space separation

To give adequate protection against the spread of fire from one building to another the provisions of Section 13 of Approved Document B4 limit the extent of unprotected area in the side of a building in relation to its distance from its relevant boundary. In this respect, any opening or any part of the wall which has less fire resistance than that shown above is considered to be an unprotected area.

Where a building is divided into fire resisting compartments by suitable floors or internal walls it is assumed that any fire will be restricted to one compartment and, therefore, the extent of unprotected area is related to such compartments only. Difficulties over unprotected areas can sometimes be resolved by upgrading the construction of a partition or floor so that it can be considered to possess the fire resistance necessary to divide the building into compartments.

Because of the particular requirements attached to the construction of a stairway which forms a protected shaft, such as would be found in a block of flats (see Chapter 15), the external wall of that stairway may be disregarded in assessing the separation distance between a building and its relevant boundary.

Also disregarded are small unprotected areas not exceeding 0.1 m^2 provided that they are no nearer to each other than 1.5 m in any direction and unprotected areas of not more than 1 m^2 (which may consist of two or more smaller areas within an area 1 m × 1 m) provided that they are at a minimum distance apart of 4 m. These disregarded areas are shown in Fig. 3.8.

If an external wall possesses the appropriate fire resistance but has a combustible facing exceeding 1 mm thick applied to it, then that part of the wall is counted as an unprotected area amounting to half the actual area of the facing.

Any wall within 1 m of its relevant boundary must be fire resisting from both sides and without any unprotected areas except for those shown in Fig. 3.8.

External walls which are more than 1 m from their relevant boundary at every point will meet the provisions for space separation if the extent of the

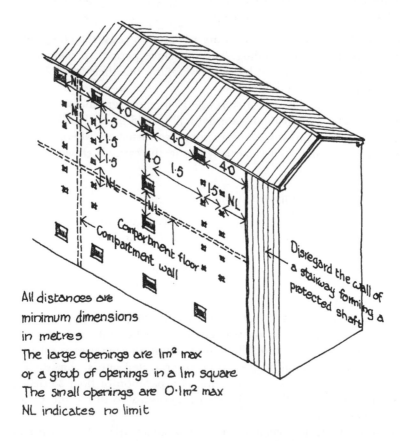

All distances are
minimum dimensions
in metres
The large openings are 1m² max
or a group of openings in a 1m square
The small openings are 0.1m² max
NL indicates no limit

Fig. 3.8 Unprotected areas which may be disregarded for separation distance purposes

unprotected areas does not exceed that given in Table 3.5 and the rest of the wall is fire resisting.

Two other methods may be used to determine the safe distance between a building and the site boundaries. The first is contained in Part 1 of a BRE Report: *External fire spread: Building separation and boundary distances*, (BRE 1991) which covers the 'Enclosing Rectangle' method, and the second is the

Table 3.5 Permitted unprotected areas

Houses and flats up to 3 storeys high and 24 m long		Residential buildings or compartments up to 10 m high	
Minimum distance from side of building to relevant boundary (m)	Maximum total sum of unprotected area (m^2)	Minimum distance from side of building to relevant boundary (m)	Maximum total percentage of unprotected areas (%)
1	5.6	1	8
2	12	2.5	20
3	18	5	40
4	24	7.5	60
5	30	10	80
6	no limit	12.5	100

(*Source:* Based on Diagram 14 and Table 16 of Approved Document B4)

Note: for any building or compartment more than 10 m in height the methods set out in the BRE Report *External Fire Spread: Building separation and boundary distance* (BRE 1991) can be applied

'Aggregate Notional Area' method to be found in Appendix J of the 1985 edition of Approved Document B2/3/4.

Chapter 4

GROUND FLOORS

4.1   The Building Regulations applicable

The Building Regulations relating to the construction of ground floors in residential buildings are:

A1   Loading
C4   Resistance to weather and ground moisture
L1   Conservation of fuel and power

A1 requires that the building, in this case the ground floor, be constructed so that the imposed loads are sustained safely; C4 states that the floors of the building must resist the passage of moisture to the inside of the building; and L1 requires that reasonable provision be made for the conservation of fuel and power.

4.2   Ground supported floors

Ground supported concrete floors comprise four elements – a hardcore base, a concrete bed, a damp-proof membrane and, in some situations, thermal insulation (see Fig. 4.1). Most floors are finished with a screed ready to receive the final flooring but this screed is not the subject of the Regulations except where the damp-proof membrane is laid on top of the concrete bed.

4.2.1 Structure

Regulation C1 requires that turf and other vegetable matter should be removed from the ground to be covered by the building at least to a depth sufficient to prevent later growth. This is usually taken to mean stripping away the surface down to the subsoil level.

The excavated level can then be made up with hardcore consisting of clean broken brick or similar inert material, free from materials including water soluble sulphates in quantities which could damage the concrete bed. No specific minimum thickness is given for this hardcore but the usual practice is to lay not less than a 100 mm depth.

On this hardcore should be laid a bed of concrete at least 100 mm thick. It may need to be thicker than this if the structural loadings are such as to require it but this is not generally the case in residential work (see Fig. 4.1). The concrete should be composed of 50 kg of cement to not more than 0.11 m^3 of sand to 0.16 m^3 of coarse aggregate. Alternatively, it can be to BS 5328 mix designation ST2.

If there is embedded steel in the floor, the concrete should be composed of 50 kg of cement to not more than 0.08 m^3 of sand to 0.13 m^3 of coarse aggregate. Alternatively, the concrete can be in accordance with BS 5328 mix ST4.

4.2.2 Damp-proofing

A damp-proof membrane must be provided within the floor construction. It can be placed either above or below the concrete bed and must be continuous with the damp-proof courses in the walls. If it is laid below the slab it should be not less than 1000 gauge polythene sheet, all joints should be sealed and the hardcore below must be finished with a bed of material which will not damage the sheet, i.e. blinded with sand.

A membrane laid above the concrete should be either polythene of 1000 or 1200 gauge or three coats of a cold applied bitumen solution or similar moisture and water vapour resisting material. In each case it should be protected by a screed or other floor finish unless the membrane is itself the floor finish such as one of the coloured pitchmastic finishes.

Alternatively, the membrane may be the adhesive bedding for a timber floor finish where it is of a type that would satisfy the requirement. If, instead of being laid in an adhesive bedding, the timber floor finish is fixed to fillets laid in the concrete, these must be treated with an effective preservative unless there is a damp-proof membrane below them (see Fig. 4.1). The Approved Document refers to BS 1282: *Guide to the choice, use and application of wood preservatives.*

The required performance of a ground supported floor can also be met by following the recommendations contained in Clause 11 of CP 102: *Protection of buildings against water from the ground.*

Where the circumstances are such that the floor will be subjected to water under pressure, the recommendation given in BS 8102 should be adopted.

4.2.3 Thermal insulation

If the Elemental Method is to be used to demonstrate compliance with the requirements in Regulation L the U value of the ground floor must be either $0.35 \text{W/m}^2\text{K}$ if the Energy Rating is 60 or less and 0.45 if the rating is over 60.

If either the Target U value Method or the Energy Rating Method is used,

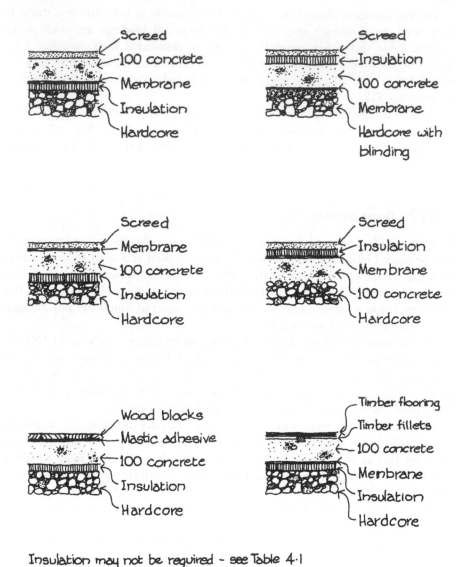

Insulation may not be required - see Table 4·1

Fig. 4.1 Ground supported floors

the thermal resistance of the ground floor can be of a lower standard provided that other elements within the fabric of the building are of a higher than minimum standard to compensate. It is recommended that, in any case the U value of any ground floor should not be worse than 0.7 W/m^{2}K, see Chapter 1, Sections 1.9 and 1.10 for an explanation of Energy Rating and the Elemental Method, the Target U value Method and the Energy Rating Methods.

To achieve these standards of thermal resistance, it is necessary, in nearly all cases, to introduce some form of insulation.

The principle that is followed in Approved Document L in assessing the required thickness of insulation is that the amount of heat loss is directly related to the size and shape of the floor. A large building with a regular shape loses less heat through the floor in relation to its size than one of an irregular shape or one of a smaller size. The reason being that with a large regularly shaped floor there is more of it remote from the cold perimeter than with an irregular or small floor. This is covered in the Approved Document by using the ratio between the perimeter length of the floor and its area as the starting point from which to determine the necessary thickness.

Table 4.1 shows the thicknesses of insulation required in three types of ground floor for a U value of 0.35 W/m^{2}K and for 0.45 W/m^{2}K. To use the table, first decide which U value is to be achieved and then find the perimeter to floor area ratio by dividing one by the other. Opposite this figure in the appropriate section of the table can be found the minimum thickness needed by reading down from the appropriate insulation material heading and floor type. This minimum thickness may need to be rounded up to the nearest greater thickness available.

This table has been based on three tables in the Approved Document and reference should be made to these for a design U value of 0.25 or for an insulating material of a thermal conductivity not shown in Table 4.1. Alternatively, the necessary thickness can be found by linear interpolation.

4.3   Suspended concrete ground floors

The use of suspended concrete ground floors has increased in recent years due no doubt to the combined effects of available economic systems and the inherent advantages of the constructional method.

4.3.1   Structure

The Approved Documents give no direct guidance on acceptable methods, materials, thicknesses or reinforcement for floors. However, the requirements of Regulation A1 will be met by following the recommendations of the British Standards listed below:

- BS 6399: *Loading for buildings*
    Part 3: *Code of practice for dead and imposed loads.*
- BS 8110: *Structural use of concrete*
    Part 1: *Code of practice for design and construction.*
    Part 2: *Code of practice for special circumstances.*
    Part 3: *Code of practice for singly reinforced beams, doubly reinforced beams and rectangular columns.*

Table 4.1 Thickness of insulation (mm) required in ground floors

Insulation material		Phenolic foam			Polyurethane board			Expanded polystyrene slab, Glass fibre slab, Mineral fibre slab			Glass fibre quilt		
Thermal conductivity W/mK		0.020			0.025			0.035			0.040		
Floor types		Solid conc.	Conc. beam	Wood joist	Solid conc.	Conc. beam	Wood joist	Solid conc.	Conc. beam	Wood joist	Solid conc.	Conc. beam	Wood joist
Design U value		0.35 W/m²K											
The length of the perimeter of the floor divided by the area of the floor	1.00	39	37	57	49	46	67	68	64	87	78	74	96
	0.90	38	36	55	48	45	66	67	63	85	76	72	94
	0.80	37	35	53	46	43	63	65	61	82	74	69	91
	0.70	35	33	51	44	41	60	62	58	78	70	66	87
	0.60	33	31	47	41	38	56	58	54	73	66	61	81
	0.50	30	27	42	37	34	50	52	48	65	59	55	72
	0.40	25	22	34	31	28	41	43	39	53	49	45	60
	0.30	16	14	22	21	18	26	29	25	35	33	29	39
	0.20	1	0	1	1	0	1	2	0	2	2	0	2
Design U value		0.45 W/m²K											
The length of the perimeter of the floor divided by the area of the floor	1.00	26	24	37	33	30	44	46	42	57	53	48	64
	0.90	25	23	35	32	29	42	44	41	55	51	46	62
	0.80	24	22	33	30	28	40	42	39	53	48	44	59
	0.70	22	20	31	28	25	37	39	36	49	45	41	54
	0.60	20	18	27	25	23	33	35	32	43	40	36	49
	0.50	17	15	22	21	18	27	30	26	36	34	29	40
	0.40	12	10	15	15	12	18	21	17	25	24	19	28
	0.30	4	2	4	5	2	5	6	3	7	7	3	8
	<0.27	0	0	0	0	0	0	0	0	0	0	0	0

(*Source:* Based on Tables A9, A10 and A11 of Approved Document L)

Notes: Floor types:
Solid conc.   ground supported in-situ concrete slab.
Conc. beam suspended concrete beam and block.
Wood joist   suspended timber joists 48 mm wide at 400 mm centres.

4.3.2 Damp-proofing

There are two requirements to be met, the floor must adequately resist moisture from reaching the upper surface and the reinforcement must be protected against moisture.

By virtue of being suspended the floor should remain dry if there is an adequate resistance to rising damp in the supporting walls. If, however, the ground beneath the floor has been excavated below the surrounding ground and will not be effectively drained then a damp-proof membrane must be provided.

To protect the steel the concrete must provide at least 40 mm of cover if the floor has been cast *in situ* and at least the thickness of cover required for 'moderate exposure' if the concrete is precast.

4.3.3 Gas ventilation

In those situations where there is a risk of gas which might lead to an explosion, accumulating under the floor, there must be a ventilated air space left below the concrete or insulation if it is provided. The space must have a clear height of at least 150 mm and be ventilated by openings in two opposing external walls so arranged that there is a free flow of air to all parts. The openings should be large enough to give an actual clear opening of not less than the equivalent of 1500 mm^2 for each metre run of wall which means one 75 × 225 mm air brick with a free area of not less than 4500 mm^2 every thirteenth stretcher (see Fig. 4.2).

4.3.4 Thermal insulation

As explained in Section 4.2.3, the thermal insulation standard for any ground floor is, normally, either 0.35 W/m^{2}K or 0.45 W/m^{2}K, depending on the Energy Rating of the building. Table 4.1 shows various insulation materials and the thicknesses needed to achieve these standards in floors of differing sizes. The thicknesses required with materials of differing thermal conductivities can be found by linear interpolation.

4.4 Suspended timber ground floors

Timber has been used for ground floors for a very long time but not always with success. The failure can nearly always be put down to a lack of observation of simple principles of construction and a lack of recognition of the fact that the material will decay in certain circumstances. The Regulation requirements represent what has been developed in the past as being good building practice.

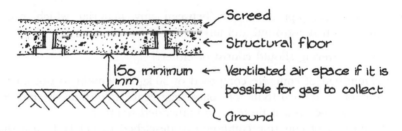

Fig. 4.2 Typical suspended concrete floor

Table 4.2 Insulation thicknesses in millimetres to achieve a *U* value of 0.45 W/m²K in suspended ground floors

Thermal conductivity of insulating material (W/mK)	Greater dimension of floor in metres					Mid-terrace building up to 10 metres from front to back
	Up to 10		10 to 15			
	Lesser dimension of floor in metres					
	Up to 10		Up to 10		10 to 15	
	Detached building	Semi-detached or end terrace	Detached building	Semi-detached or end terrace	Detached building	
0.025	26	22	22	19	12	16
0.03	31	26	26	23	14	20
0.035	36	30	30	27	17	23
0.04	42	34	34	30	19	26
0.045	47	39	39	34	21	29
0.05	52	43	43	38	24	33

(*Source:* Based on Table 7 of Approved Document L1)

Notes: 1. Where the estimated thickness is not available the next greater thickness should be used.
2. Where the thermal conductivity of a selected material is not shown an interpolated reading may be used.
3. No insulation is required in semi-detached or end terrace buildings where the floor dimensions exceed 10 metres in both directions.
4. No insulation is needed in mid-terrace buildings over 10 metres deep.
5. There is no simple calculation available as an alternative to this table.

4.4.1 Structure

A timber ground floor will meet the performance requirements if:

- the ground is covered so as to resist moisture and prevent the growth of plants below the floor;
- there is a ventilated air space between the ground covering and the timber structure;
- there is a damp-proof course between any material which can carry moisture from the ground and any timber.

The ground covering should consist of either:

1 a 100 mm bed of concrete composed of 50 kg of cement to not more than 0.13 m³ of sand to 0.18 m³ of coarse aggregate (alternatively, mix ST1 of BS 5328 can be used if there is no embedded steel); or
2 a 50 mm bed of concrete composed as described in (1) laid on a sheet of polythene of minimum 1000 gauge. The polythene must have the joints

sealed and be laid on a bed of material which will prevent damage to the sheet.

To prevent water collecting below the floor and on top of the ground covering it should be laid so that its upper surface is above the level of the adjoining ground or, if this is not feasible, laid to falls with a drainage outlet through the wall above the lowest level of the ground (see Fig. 4.3).

The ventilated air space should measure at least 75 mm from the ground covering to the underside of any wall plates and at least 150 mm to the underside of the floor joists (or insulation if it is provided and lower than the joists). The ventilation is to be provided by openings placed in two opposite walls so that there will be a free flow of air to all parts. The openings should be large enough to give a free area of not less than the equivalent of 1500 mm^2 for every metre run of wall, or, in other words, a 75 × 225 mm air brick (with a free area of 4500 mm^2 or more) every thirteenth stretcher.

Any pipes required to duct the ventilating air to the underfloor space must have a minimum diameter of 100 mm.

The damp-proof course should consist of an impervious sheet material, engineering bricks in cement mortar, slates in cement mortar or any other material that will prevent the passage of moisture. It is usual to lay 50 or 75 × 100 mm timber wall plates on the brickwork provided to support the floor joists and, if so, the damp-proof coursing should be placed immediately below these plates (see Fig. 4.4). Note that the joists should bear at least 35 mm on to the plates.

The sleeper walls, provided to support the wall plates, should be honeycombed so that they do not prevent the free flow of ventilating air required and can be built off the ground covering described in (1) above, usually at approximately 1200 to 1800 mm centres.

The floor joists are fixed at 400 mm centres to coincide with the edges of 1200 mm flooring panels or, if floor boards are used, the spacing can be

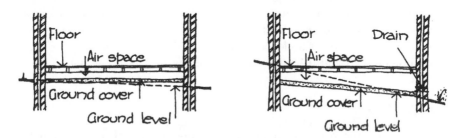

Fig. 4.3 Prevention of water collecting under suspended timber ground floors

increased to 450 or 600 mm. The size of the joists depends on the spacing and the span and can be found from Table 4.3 which is extracted from Tables A1 and A2 of Appendix A of Approved Document A1/2. Table 4.3 shows just the timber sizes which are relevant to ground floors, the full range of sizes can be found in Chapter 5.

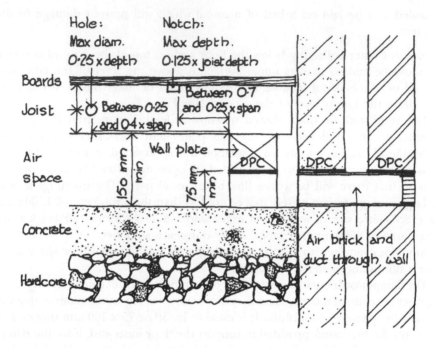

Fig. 4.4 Suspended timber ground floor (floor insulation omitted for clarity)

4.4.2 Notches and holes

Cutting away the joists for services is permitted but only within the following limits:

1 notches should not be deeper than 0.125 times the depth of the joist and their position should be at a distance from the support of between 0.07 and 0.25 times the span of the joist;
2 holes should not be greater than 0.25 times the depth of the joist in diameter, drilled at mid-depth and not closer together than three times their diameter. They should be located at a distance from the joist support of between 0.25 and 0.4 times the span of the joist (see Fig. 4.4).

4.4.3 Flooring

Appendix A of Approved Document A1/2 refers only to softwood tongued and grooved floorboards as a finish to be laid over the joists and stipulates that, for joist spacings of 400 mm and 450 mm, the boards should be not less than 16 mm finished thickness. Wider spacings, up to 600 mm, require a finished thickness of 19 mm.

Flooring grade particle board panels are recommended to be 18 mm thick for joist spacings of 400 mm and 450 mm and 22 mm thick for spacings of 600 mm.

Table 4.3 Spans of timber joists in suspended ground floors

Size of joist	Timber strength class SC3			Timber strength class SC4		
	Maximum clear span in metres for dead loads not exceeding 0.25 kN/m^2			Maximum clear span in metres for dead loads not exceeding 0.25 kN/m^2		
	Spacing in millimetres			Spacing in millimetres		
	400	450	600	400	450	600
38 × 97	1.83	1.69	1.30	1.94	1.83	1.59
38 × 122	2.48	2.39	1.93	2.58	2.48	2.20
38 × 147	2.98	2.87	2.51	3.10	2.98	2.71
47 × 97	2.02	1.91	1.58	2.14	2.03	1.76
47 × 122	2.66	2.56	2.30	2.77	2.66	2.42
47 × 147	3.20	3.08	2.79	3.33	3.20	2.91
50 × 97	2.08	1.97	1.67	2.20	2.09	1.82
50 × 122	2.72	2.62	2.37	2.83	2.72	2.47
50 × 147	3.27	3.14	2.86	3.39	3.27	2.97

(*Source:* Extracted from Tables A1 and A2 of Appendix A of Approved Document A1/2)

Notes: 1. Joist spacing is the dimension centre to centre of the timbers.
2. Span is the clear span between supports.
3. The dead load assumes softwood floorboards or particle board flooring panels.
4. The imposed load must not exceed 1.5 kN/m^2, double joists are required beneath a bath.

4.4.4 Thermal insulation

The requirements for the insulation of suspended timber floors follow those of ground supported floors and suspended concrete floors given in Sections 4.2.3 and 4.3.4. Table 4.1 shows the thicknesses required for insulation placed between the floor joists. It is assumed in the Table that the timber represents 12 per cent of the area of the floor which is equivalent to 48 mm wide joists at 400 mm centres. For other proportions of timber, the *U* value can be calculated using the procedure outlined in Appendix B of Approved Document L.

Chapter 5

SUSPENDED UPPER FLOORS

5.1 The Building Regulations applicable

The Building Regulations which apply to the construction of suspended floors in domestic work are:

A1 Loading
B2 Internal fire spread (linings)
B3 Internal fire spread (structure)
E2 Airborne sound (floors and stairs)
E3 Impact sound (floors and stairs)
L1 Conservation of fuel and power

A1 requires that the building, in this case the upper floors, be constructed so that the combined dead, imposed and wind loads are sustained safely; B2 states that the internal linings of a building, in this respect the ceilings, must resist the spread of flame over their surface and, if ignited, have a rate of heat release which is reasonable in the circumstances; B3 has two requirements in relation to upper floors, firstly, that the building shall be designed and constructed so that, in the event of fire, its stability will be maintained for a reasonable period and, secondly, to inhibit the spread of fire within the building, it must be subdivided with fire resisting construction to an extent appropriate to the size and intended use of the building; E2 and E3 stipulate that the floor separating a dwelling from another dwelling or other part of the building which is not part of the dwelling must resist the passage of airborne sound and impact sound, respectively; and L1 requires reasonable provision to be made to conserve fuel and power, in this application, by insulating floors exposed to the outside or unheated spaces.

5.1.1 Compartmentation

The principle of dividing the building into compartments is frequently referred to in connection with the control of fire. Approved Document B states:

> The spread of fire within a building can be restricted by sub-dividing it into compartments separated from one another by walls and/or floors of fire resisting construction. The object is two-fold:

a. to prevent rapid fire spread which could trap occupants of the building; and
b. to reduce the chance of fires becoming large, on the basis that large fires are more dangerous, not only to occupants and fire service personnel but to people in the vicinity of the building.

In the case of a house the only compartment floor required is the first floor over an integral garage. With flats and maisonettes, all floors separating parts in different occupancies must be compartment floors with a fire resistance as shown in Section 5.2.5. Floors between storeys within a maisonette are not required to be of compartment standard.

5.2   Suspended timber floors

Timber is still the most popular material for the suspended floors of houses where the spans are small enough for it to be the most economic material from the aspect of strength/weight ratio and where there are no requirements for sound insulation or fire resistance. In the construction of flats the spans are also relatively small but the sound insulating requirements and the necessary fire resistance to be built into the floor make concrete a more attractive proposition, although, as shown in the following sections, it is possible to meet the Building Regulation requirements using timber. Maisonettes having separating floors and internal floors may be built with a mixture, for example concrete floors between maisonettes and timber floors between the storeys within the maisonette.

5.2.1   Structure

The size of the structural timbers used as joists in a suspended timber floor is determined by the strength of the timber, the load to be carried, the span of the floor and the spacing of the timbers.

For the purposes of the Building Regulations, there are two strength classes relating to timber: SC3 and SC4. Strength Class SC3 generally covers softwoods such as imported redwood and whitewood, homegrown fir, larch, pine and spruce as well as those same timbers from Canada and the USA all to grade 'general selected' (GS or MGS) or M50. Strength Class SC4 requires the same timbers to be to a grading of special selected (SS or MSS). Details of the timber species, their origin, the appropriate grading rules and the grades are to be found in Table 1 of Section 1B of Approved Document A1/2 or in BS 5268: Part 2: 1991.

The dead loads to be allowed for in selecting a joist size are shown in the tables of joist sizes given in Table 5.1. They are due to the floor covering and the ceiling to be carried. Typical loads for flooring and ceiling materials are as follows:

Materials	Load (kN/m^2)
Flooring only	0.08 to 0.11
Fooring and plasterboard and skim ceiling	0.23 to 0.26
Fooring and two coat plaster ceiling	0.27 to 0.35
Fooring with parquet finish and two coat plaster ceiling	0.32 to 0.42

Table 5.1   Spans of timber joists in suspended upper floors

Size of joist	Timber strength class SC3						Timber strength class SC4					
	Maximum clear span in metres for dead loads not exceeding:						Maximum clear span in metres for dead loads not exceeding:					
	under 0.25 kN/m²			0.25 to 0.5 kN/m²			under 0.25 kN/m²			0.25 to 0.5 kN/m²		
	Spacing in millimetres			Spacing in millimetres			Spacing in millimetres			Spacing in millimetres		
	400	450	600	400	450	600	400	450	600	400	450	600
38×97	1.83	1.69	1.30	1.72	1.56	1.21	1.94	1.83	1.59	1.84	1.74	1.51
38×122	2.48	2.39	1.93	2.37	2.22	1.76	2.58	2.48	2.20	2.47	2.37	2.08
38×147	2.98	2.87	2.51	2.85	2.71	2.33	3.10	2.98	2.71	2.97	2.85	2.59
38×170	3.44	3.31	2.87	3.28	3.10	2.69	3.58	3.44	3.13	3.43	3.29	2.99
38×195	3.94	3.75	3.26	3.72	3.52	3.06	4.10	3.94	3.58	3.92	3.77	3.42
38×220	4.43	4.19	3.65	4.16	3.93	3.42	4.61	4.44	4.03	4.41	4.25	3.86
47×97	2.02	1.91	1.58	1.92	1.82	1.46	2.14	2.03	1.76	2.03	1.92	1.68
47×122	2.66	2.56	2.30	2.55	2.45	2.09	2.77	2.66	2.42	2.65	2.55	2.29
47×147	3.2	3.08	2.79	3.06	2.95	2.61	3.33	3.20	2.91	3.19	3.06	2.78
47×170	3.69	3.55	3.19	3.53	3.40	2.99	3.84	3.69	3.36	3.67	3.54	3.21
47×195	4.22	4.06	3.62	4.04	3.89	3.39	4.39	4.22	3.85	4.20	4.05	3.68
47×220	4.72	4.57	4.04	4.55	4.35	3.79	4.86	4.73	4.33	4.71	4.55	4.14
50×97	2.08	1.97	1.67	1.98	1.87	1.54	2.20	2.09	1.82	2.08	1.98	1.73
50×122	2.72	2.62	2.37	2.60	2.50	2.19	2.83	2.72	2.47	2.71	2.60	2.36
50×147	3.27	3.14	2.86	3.13	3.01	2.69	3.39	3.27	2.97	3.25	3.13	2.84
50×170	3.77	3.62	3.29	3.61	3.47	3.08	3.91	3.77	3.43	3.75	3.61	3.28
50×195	4.31	4.15	3.73	4.13	3.97	3.50	4.47	4.31	3.92	4.29	4.13	3.75
50×220	4.79	4.66	4.17	4.64	4.47	3.91	4.93	4.80	4.42	4.78	4.64	4.23
75×122	3.10	2.99	2.72	2.97	2.86	2.60	3.22	3.10	2.83	3.09	2.97	2.71
75×147	3.72	3.58	3.27	3.56	3.43	3.13	3.86	3.72	3.39	3.70	3.57	3.25
75×170	4.28	4.13	3.77	4.11	3.96	3.61	4.45	4.29	3.91	4.27	4.11	3.75
75×195	4.83	4.70	4.31	4.68	4.52	4.13	4.97	4.83	4.47	4.82	4.69	4.29
75×220	5.27	5.13	4.79	5.11	4.97	4.64	5.42	5.27	4.93	5.25	5.11	4.78

(*Source:* Based on Tables A1 and A2 of Appendix A to Approved Document A1/2)

The sizes, spacing and spans will support the dead loads shown in the table plus an imposed load not exceeding $1.5 \, \text{kN/m}^2$.

The loads due to partitions built off the floor are not allowed for in the table and special provision must be made where these occur.

Where a bath is to be installed the joists which will support it should be doubled.

Notching of the joists is allowed within the limits set out in Section 5.2.3 and the joists should be strutted as described in Section 5.2.2.

Table 5.1 includes just those timber sizes which are more commonly available; for the full range of sizes and for dead loading of between 0.5 and $1.25 \, \text{kN/m}^2$, reference should be made to Tables A1 and A2 of Approved Document A1/2.

5.2.2 Strutting

Joists spanning less than 2.5 m do not need to be strutted, those spanning between 2.5 and 4.5 m require one row of strutting at mid-span and for those over 4.5 m span, two rows are needed equally spaced along the length of the joist.

Solid strutting should be of timber at least 38 mm thick and not less than three-quarters of the depth of the joist. Herringbone strutting should be of timber at least 38×38 mm fixed as shown in Fig. 5.1. This form of strutting is not suitable where the distance between the joists is greater than three times their depth.

5.2.3 Notches and holes

Notches and holes in joists must be restricted to the following limits (see Fig. 5.2):

1 Notches are to be no deeper than 0.125 times the depth of the joist.
2 Notches cut in the joist must be positioned towards one of the ends at a distance of between 0.7 and 0.25 times the span away from the support.
3 Holes should not have a larger diameter than one quarter of the depth of the joist.
4 Holes should be drilled in the middle of the depth of the joist, at least three times the diameter apart and at a distance from the support of between 0.25 and 0.4 times the span.

5.2.4 Flooring

The Approved Document refers only to softwood tongued and grooved floorboards and requires that for joist spacings up to 500 mm the finished thickness of the flooring should be not less than 16 mm and for spacings up to 600 mm it is to be 19 mm.

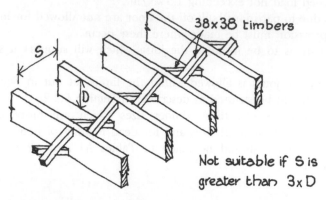

38x38 timbers

S

D

Not suitable if S is
greater than 3 x D

HERRINGBONE STRUTTING

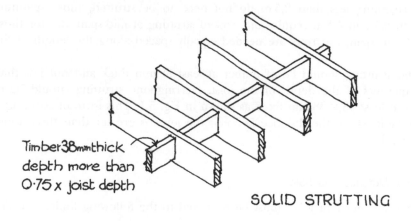

Timber 38mm thick
depth more than
0.75 x joist depth

SOLID STRUTTING

No strutting required under 2.5m span
1 row required 2.5m to 4.5m span
2 rows at 1/3rd points over 4.5m span

Fig. 5.1 Strutting of floor joists

Manufacturers recommend that tongued and grooved chipboard flooring panels should be not less than 18 mm thick for joist spacings of 400 mm and 450 mm and 22 mm thick for spacings of 600 mm.

5.2.5 Fire resistance

The fire resistance of a floor (and any other element of structure) is expressed as a time and represents a measure of its ability to withstand the effects of fire in one or more of the following ways:

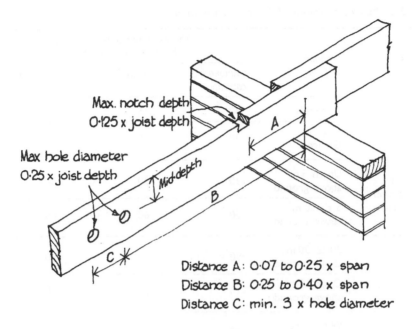

Max. notch depth
0·125 x joist depth

Max hole diameter
0·25 x joist depth

Mid depth

A

B

C

Distance A: 0·07 to 0·25 x span
Distance B: 0·25 to 0·40 x span
Distance C: min. 3 x hole diameter

Fig. 5.2 Notches and holes in joists

1 *Loadbearing capacity* This is applicable to a structural element only and indicates its resistance to collapse.
2 *Integrity* This refers to the capacity of the construction to prevent the fire breaking through from one room to another.
3 *Insulation* This is the ability of the floor to resist the transference of an excessive amount of heat from one side to the other.

The minimum periods of resistance required in the floors of houses and flats when exposed to fire from below are shown in Table 5.2.

The Approved Documents do not give any guidance on ways of achieving these standards but refer to specifications given in Part II of the Building Research Establishment's Report *Guidelines for the construction of fire resisting structural elements* (BRE 1988). The following notes are taken from this publication.

Table 5.3 shows various constructions and their appropriate fire resistance period. The term 'modified half hour' refers to the special requirement applicable to two storey houses and shown in Table 5.2 where the construction must be capable of carrying the loads applied for not less than 30 minutes but its resistance to flames breaking through or its insulation value being lost is only 15 minutes. The table is based on joists of a width of 37 mm. If this is increased it may be possible to reduce the ceiling specification but guidance on this should be sought from the manufacturer concerned. The actual joist size must be based on the load to be carried as shown in Table 5.1 but with the condition that, for fire resistance purposes, the maximum spacing for 9 mm

Table 5.2 Minimum periods of fire resistance for upper floors

Building use	Location of floor	The minimum time (in minutes) the floor must maintain its:		
		Loadbearing capacity	Integrity	Insulation
Two storey houses	Generally	30	15	15
	Over a Garage	30	30	30
Flats and Houses over two storeys	Up to 5 m above ground	30	30	30
	5 to 20 m above ground	60	60	60
Flats	20 to 30 m above ground	90	90	90
	Over 30 m above ground	120	120	120
Maisonettes	Within the maisonette*	30	30	30

(*Source:* extracted from Diagram 22 and Tables A2 and A3 of Appendix A of Approved Document B)

* Floors separating maisonettes must have the same standards as those separating flats.

plasterboard should be 450 mm but for 12.5 mm plasterboard it can be 600 mm.

Ceiling boards must be fixed to every joist and heading joints fixed to noggins. In double layer constructions, each layer must be fixed independently. Fixings are to be with galvanized nails, of the sizes shown below, at 150 mm centres:

- 9.5 mm plasterboard use 30 mm nails;
- 12.5 mm plasterboard use 40 mm nails;
- 19 mm to 25 mm plasterboard use 60 mm nails.

Expanded metal lath should be fixed at 100 mm centres with 38 mm nails or 32 mm staples. End laps should be not less than 50 mm, side laps not less than 25 mm and both wired together at 150 mm intervals. The lath should be spaced away from the background to provide a 6 mm gap to ensure adequate mechanical bond for the plaster.

5.2.6 Sound insulation

The floors of any building with more than one occupancy must resist the passage of sound generated by either a sound source such as a radio (airborne

Table 5.3 Fire resisting suspended timber floors

Construction and materials	Floor finish of 15 mm minimum				Floor finish of 21 mm minimum			
	Minimum thickness (mm) of protection for a fire resistance of:							
37 mm Joists (min) with a ceiling of:	Modified half hour	Half hour	One hour	Two hours	Modified half hour	Half hour	One hour	Two hours
Metal lathing and sanded gypsum plaster of:		15				15		
Metal lathing and lightweight aggregate gypsum plaster of:		13	13	25*		13	13	25*
One layer of plasterboard with taped and filled joints:	12.7				12.7	12.7		
Two layers of plasterboard with joints staggered, joints in outer layer taped and filled total thickness:	19	22	31†		19		31†	
One layer of 9.0 mm plasterboard and gypsum plaster finish of:	5				5			
One layer of 9.0 mm plasterboard and sanded gypsum plaster finish of:		15				13		
One layer of 9.0 mm plasterboard and lightweight aggregate gypsum plaster (Mix V) finish of:		13				13		
One layer of 12.5 mm plasterboard and gypsum plaster of:		5				5		
One layer of 12.5 mm plasterboard and lightweight aggregate gypsum plaster (Mix V) finish of:		10				10		

(Source: Based on Table 14 of the BRE Report: Guidelines for the construction of fire resisting structural elements)

Notes: The floor finishes referred to in the table are: tongued and grooved boarding or tongued and grooved plywood or wood chipboard.
* Metal lathing to be also independently fixed with wire supports from the joist sides.
† In conjunction with 50 mm (min) wide joists and plywood or chipboard flooring.

sound) or an object hitting the structure (impact sound). The means of providing resistance to each is different. Airborne sound is absorbed by the mass of the floor, impact sound is reduced at source by a resilient floor surface. Since timber floors are a light form of construction, insulation against the direct transmission of sound is more difficult than is the case with concrete floors, none the less it is possible by the addition of pugging to increase the mass of the construction.

As well as the direct transmission of sound it is necessary to prevent

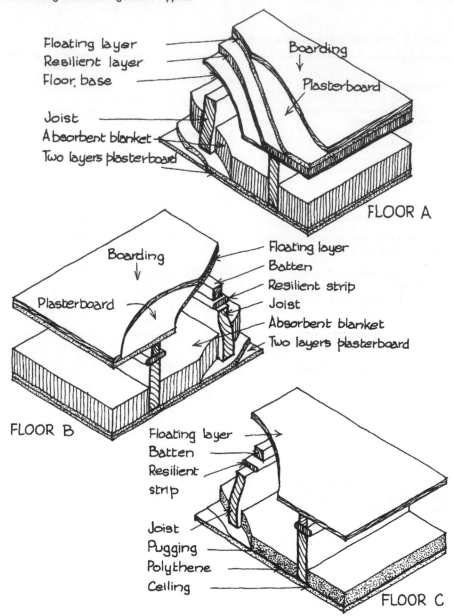

Floating layer
Resilient layer
Floor base

Joist
Absorbent blanket
Two layers plasterboard

Boarding

Plasterboard

FLOOR A

Boarding

Plasterboard

Floating layer
Batten
Resilient strip
Joist
Absorbent blanket
Two layers plasterboard

FLOOR B

Floating layer
Batten
Resilient strip

Joist
Pugging
Polythene
Ceiling

FLOOR C

Fig. 5.3 Sound insulating suspended timber floors

flanking transmission round the edges of the floor and transmission through air gaps such as could occur where pipes penetrate the structure.

Approved Document E defines three timber floor constructions considered to be satisfactory. These are reproduced in Fig. 5.3 and described in more detail below.

Floor construction A (Fig. 5.3) is a platform floor with absorbent material constructed with the following:

1 Structural joists of the size required for the span, spacing and load.
2 A floor base of 12 mm softwood boarding or wood-based board.
3 A resilient layer of mineral fibre, 25 mm thick with a density of between 60 and 100 kg/m^3.
4 A floating layer of either:
   (a) Tongued and grooved timber or wood-based boarding 18 mm thick, all joints glued and spot bonded to a sublayer of 19 mm plasterboard; or
   (b) two thicknesses of cement bonded particle board, glued and screwed together with the joints staggered giving a total thickness of 24 mm.
5 A ceiling of two layers of plasterboard giving a total thickness of 30 mm, fixed to the underside of the joists with the joints staggered.
6 An absorbent layer of unfaced mineral wool of a density of not less than 10 kg/m^3, 100 mm thick, laid on the ceiling.

Floor construction B is a ribbed floor with absorbent material constructed with the following:

1 Structural joists of the required size but not less than 45 mm wide.
2 Resilient strips of mineral fibre of a density of 80 to 140 kg/m^3 and 25 mm thick laid along the joists.
3 45 × 45 mm battens laid on top of the resilient strips.
4 A floating layer forming the floor surface of tongued and grooved softwood or wood-based boarding 18 mm thick, all joints glued and spot bonded to an underlayer of plasterboard 19 mm thick nailed to the battens only.
5 A ceiling of two layers of plasterboard giving a total thickness of 30 mm nailed to the underside of the joists with the joints staggered.
6 An absorbent blanket of unfaced rock fibre of minimum density of 10 kg/m^3 and 100 mm thick laid on the ceiling.

There are proprietary products which achieve the same result as the floating layer given, indeed, can achieve a more superior result. These comprise a double layer of gypsum fibreboard with a backing of polystyrene foam or mineral wool and range in thickness from 20 mm to 65 mm.
The mineral wool backing provides acoustic and fire insulation.

Floor construction C is a ribbed floor with heavy pugging constructed with the following:

1 Floor joists of the required size for the span, spacing and load but not less than 45 mm wide.
2 Resilient strips of mineral fibre of a density of 80 to 140 kg/m^3, laid along the joists.
3 45 × 45 mm timber battens laid on the resilient strips.
4 A finished floor of softwood or wood-based boards, 18 mm thick, all joints glued and nailed to the battens only.
5 A ceiling of either:
   (a) dense plaster 19 mm thick on expanded metal lath; or
   (b) plywood 6 mm thick plus two layers of plasterboard giving a total thickness of 25 mm fixed with the joints staggered.
6 Pugging, laid on polythene on the ceiling consisting of:
   a bed of ash, 75 mm thick, or a bed of dry sand, 50 mm thick, or a 60 mm bed of limestone chippings or whin aggregate, graded 2 to 10 mm.

Sand should not be used in situations such as bathroom or kitchen floors where it may become wet and overload the structure.

Because of the number of parts to these floor constructions and the mass involved, the floor joist sizes given in Table 5.1 are not suitable and the required joist size should be calculated to take account of the dead loads involved.

To limit flanking transmission through the walls round the perimeter of the floor, the floating layer should be isolated from the face of the wall with a resilient strip (in the case of floor A the resilient layer can be turned up the wall) and there should be a 3 mm gap below the skirting board. In addition, the junction between the ceiling and the wall should be sealed with tape or caulked.

Any sound insulating constructions differing from those given, which have already been built, tested as specified in Approved Document E, BS 2750: Part 4: 1980 and BS 2750: Part 7: 1980 and which have been shown to be satisfactory, can be repeated in other new work.

5.2.7 Thermal insulation

Two heat loss situations are taken into account in Approved Document L and are referred to as exposed and semi-exposed floors. Exposed floors are those where the soffite of the floor is exposed to external conditions and temperatures; semi-exposed floors are those which are placed over an unheated and uninsulated space such as a garage. In addition, two standards are set, depending on the Energy Rating of the house or flat, as follows:

	SAP Rating 60 or less	SAP Rating over 60
Exposed floors	0.35 W/m^{2}K	0.45 W/m^{2}K
Semi-exposed floors	0.60 W/m^{2}K	0.60 W/m^{2}K

(see Chapter 1 for an explanation of Energy Rating and Chapter 3 for an explanation of U value).

Table 5.5 gives the calculated thickness required to achieve these thermal insulation standards. The thicknesses taken from the Table are theoretical and will have to be rounded up to the next greater thickness commercially available. Thicknesses for material with a thermal conductivity other than those shown can be found by linear interpolation.

5.3 Suspended concrete floors

Although rarely used in houses, concrete floors are frequently employed in blocks of flats because the standards of fire resistance and sound insulation are far more easily achieved.

5.3.1 Structure

Since a concrete floor is either purpose designed for its location or a proprietary system using prefabricated units there are no specific guidelines in any Approved Documents as to their construction but Approved Document A1/2 lists the following British Standards as being relevant:

- BS 6399: *Loading for buildings*
  Part 1: *Code of practice for dead and imposed loads.*
- BS 8110: *Structural use of concrete*
  Part 1: *Code of practice for design and construction.*
  Part 2: *Code of practice for special circumstances.*
  Part 3: *Code of practice for singly reinforced beams, doubly reinforced beams and rectangular columns.*

5.3.2 Fire resistance

The standards of fire resistance required to be achieved in a concrete floor depend on the use of the building and the height of the floor above ground level. The precise requirements are shown in Table 5.2.

No specific guidance is given in any Approved Document as to how to achieve these standards in a concrete floor but Approved Document requires that it should conform to an appropriate specification in Part II of the Building Research Establishment's Report: *Guidelines for the construction of fire resisting structural elements* (BRE 1988). The following notes are taken from this publication.

The fire resistance properties of a concrete floor depend on the thickness of the slab and the thickness of the cover to the reinforcement, i.e. the distance from the heated face to the steel. The need for a certain slab thickness is obvious, the need to provide a minimum amount of concrete around the reinforcement is to ensure that the heat does not cause it to spall off thereby reducing the grip of the steel in the slab and weakening the strength of the structure.

Table 5.4 shows the thicknesses to be provided for different periods of fire resistance of a concrete floor with a flat soffit. The slab thicknesses given in the table relate to solid floors, if it is a hollow slab or beam and block construction an effective thickness must be used instead, calculated from the actual thickness and the proportion of material per unit width. In these cases the manufacturer should be able to provide the necessary information.

5.3.3 Sound insulation

As stated above, any floor dividing a building in separate occupancies must be insulated against airborne and impact sound. Being of a greater mass than a timber floor, a concrete floor possesses an inherent sound insulating value because the sound energy is absorbed by the greater mass. It may still be necessary to make special provisions to improve this inherent value to meet the requirements of the Regulations and Fig. 5.4 shows two floor constructions that are considered satisfactory.

Table 5.4 Fire resisting *in situ* suspended concrete floors

Construction and materials		Minimum dimensions (mm) excluding any finish, for a fire resistance of:			
		30 min	60 min	90 min	120 min
Reinforced concrete, simply supported, using					
(a) dense concrete	thickness	75	95	110	125
	cover	15	20	25	35
(b) lightweight concrete	thickness	70	90	105	115
	cover	15	15	20	25
Reinforced concrete, continuous, using					
(a) dense concrete	thickness	75	95	110	125
	cover	15	20	20	25
(b) lightweight concrete	thickness	70	90	105	115
	cover	15	15	20	20

(Source: Based on Table 10 of BRE Report *Guidelines for the construction of fire resisting structural elements*)

Floor type 1 (Floor A in Fig. 5.4) is composed of a structural slab with a resilient floor finish applied to it. The structural slab, including any screeding, permanent shuttering or ceiling finish, is to have a mass of not less than 365 kg/m^2 and the resilient finish is to have an uncompressed thickness of not less than 4.5 mm. The disadvantage of this method is its reliance on the floor finish to absorb impact sound. When new it will be satisfactory but after some time it will wear thin, thereby reducing its effect, or it may be replaced by another, non-resilient finish which will deprive the floor of its resistance to impact sound altogether.

Floor type 2 (Floor B in Fig. 5.4) does not have this disadvantage because the floating layer is not the floor finish and, therefore, any finish may be laid and subsequently relaid without affecting the sound insulation performance. The structure of the floor is as for floor type 1 but its total mass need only be 300 kg/m^2.

The floating layer can be either tongued and grooved timber or wood-based boarding fixed to 45 × 45 mm battens or a cement and sand screed 65 mm thick with wire mesh reinforcement. Either type of floating layer can be laid on to a resilient layer of mineral fibre of a density of 36 kg/m^3 and 25 mm thick. If the battens of the timber floating layer have an integral closed cell resilient

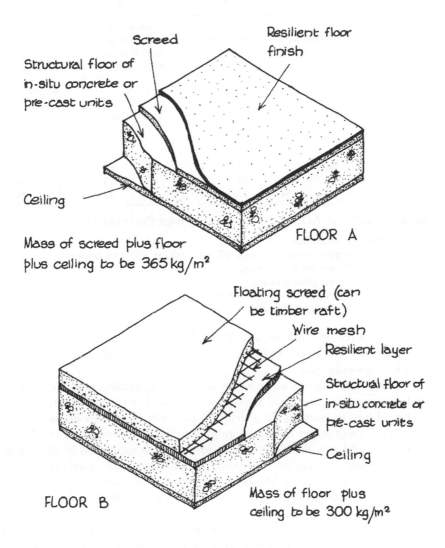

Screed

Resilient floor finish

Structural floor of in-situ concrete or pre-cast units

Ceiling

Mass of screed plus floor plus ceiling to be 365 kg/m²

FLOOR A

Floating screed (can be timber raft)

Wire mesh

Resilient layer

Structural floor of in-situ concrete or pre-cast units

Ceiling

FLOOR B

Mass of floor plus ceiling to be 300 kg/m²

Fig. 5.4 Sound insulating suspended concrete floors

foam strip, the thickness of the mineral fibre can be reduced to 13 mm. The fibre should be paper faced and laid paper face downwards under a timber finish and paper upwards under a screed.

Screeded floating layers can also be laid over pre-compressed expanded polystyrene board, 13 mm thick, of the impact sound duty grade or extruded (closed cell) polyethylene foam of a density of 30 to 45 kg/m³ and 5 mm thick.

The mineral fibre and the expanded polystyrene should be laid tightly butted and the polyethylene foam should have the joints lapped. In all cases the insulant should be turned up at the edges to isolate the floating layer from

Table 5.5 Thickness of insulation (mm) required in suspended upper floors

Insulation material	Phenolic foam board		Polyurethane board		Expanded polystyrene slab, Glass fibre slab, Mineral fibre slab		Glass fibre quilt	
Thermal conductivity	0.020 W/mK		0.025 W/mK		0.035 W/mK		0.040 W/mK	
Floor types	Timber	Concrete	Timber	Concrete	Timber	Concrete	Timber	Concrete
Design U value W/m²K								
0.35	56	50	71	62	99	87	114	99
0.45	38	37	48	46	67	65	76	74
0.60	21	24	27	30	37	42	42	48

(*Source:* Based on Tables A12, A13 and A14 of Approved Document L)

Notes: Floor types specification:

  Timber: Suspended timber joists, 48 mm wide at 400 mm centres,
        19 mm timber flooring,
        13 mm plasterboard.

  Concrete: Suspended concrete floor of any type,
        50 mm screed finish.

the enclosing walls and the skirting board left with a small gap between its bottom edge and the floor. Flanking transmission with both types of floor is dealt with in a similar way.

Where the floor adjoins an external or cavity separating wall the mass of the wall leaf next to the floor should be at least $120 \, \text{kg/m}^2$ unless it is an external wall having openings of at least 20 per cent of its area in each room, in which case there is no minimum requirement. The floor base (excluding any screed) should pass through the leaf, whether it is supported by the leaf or not but it should not bridge the cavity. If the floor base is constructed of concrete beams with infilling blocks or concrete planks, the first joint should be not less than 300 mm from the faces of the walls parallel to the structural elements.

Where the floor meets an internal partition wall with a mass of less than $375 \, \text{kg/m}^2$, the floor base, excluding any screeds, should pass through the wall. If the wall density is more than $375 \, \text{kg/m}^2$, it can pass through the floor and be tied to it with the joint grouted or, alternatively, the floor can pass through the wall.

5.3.4 Thermal insulation

The three thermal insulation standards given in Section 5.2.7 also apply to concrete floors and Table 5.5 shows the thickness of insulation required to achieve them. This theoretical thickness can be rounded up to the next greater thickness commercially available.

5.4   Suspended ceilings

Ceilings suspended below floors provide a ready route for the spread of smoke and flame through the cavity thus created. To protect against this hazard, the cavity should be fitted with a fire resisting barrier to coincide with any other fire resisting partitions within the building which divide the interior into compartments. This is dealt with in more detail in Chapter 7.

Alternatively, the ceiling itself may be fire resisting. For this purpose, the ceiling should:

1 have a spread of flame rating for its upper surface, i.e. facing the cavity of Class 1;
2 have a spread of flame rating for its soffit of Class O;
3 have a fire resistance of at least 30 minutes;
4 be imperforate except for the passage of pipes (which should be properly fire stopped – see Section 5.5), fire resisting ducts or ducts fitted with automatic fire shutters where they pass through the ceiling, cables or conduits containing one or more cables;
5 extend throughout the building or compartment;
6 not be demountable.

5.5   Pipes and ducts through floors

The necessity to provide protection where the passage of a pipe or duct through a floor creates an opening only arises where the floor is required to possess fire resisting or sound insulating properties. In many situations, the floor must satisfy both criteria, principally where it is a compartment floor in a block of flats.

All pipes, except gas pipes, which penetrate a sound insulating floor are to be enclosed in a duct above and below the floor. The duct is to be constructed with:

1 material of a mass of at least 15 kg/m^2;
2 internal lining with unfaced mineral wool 25 mm thick, or, alternatively, the pipe may be wrapped with this;
3 a 3 mm gap between the bottom of the duct and the floating floor surface, sealed with acrylic caulking or neoprene;
4 flexible fire stopping between the pipe and the edges of the hole in the floor so arranged as to prevent any rigid contact between them.

In the Gas Safety Regulations there are provisions for gas pipes to be in separate, ventilated ducts or they can remain unducted.

Pipes penetrating fire resisting floors should be sealed with a proprietary fire seal, fire stopped, sleeved or encased in a duct. With a proprietary seal, the pipe can be of any diameter but if fire stopping is used the nominal internal diameter should not exceed:

- 160 mm for a pipe in any non-combustible material;
- 160/110 mm for a stack pipe/branch pipe in lead, aluminium alloy, PVC or fibre cement and encased in a fire resisting duct;
- 40 mm for a stack pipe or branch pipe as above but not in a duct;
- 40 mm for a pipe in any material other than those above.

The fire stopping may be a proprietary material or cement mortar; gypsum based plaster; cement or gypsum based vermiculite/perlite mixes; glass fibre, crushed rock, blast furnace slag or ceramic based products (with or without resin binders), and intumescent mastics.

A pipe in lead, aluminium alloy, PVC or fibre cement may be run in a sleeve pipe through the floor which must project 1 m above and below the floor surface or soffit, be of non-combustible material and be in contact with the service pipe.

Where the pipe is to be encased in a fire resisting duct all the internal surfaces, including the bounding wall faces, are to have a spread of flame classification of class O. The casing itself should:

1 have a fire resistance of not less than 30 minutes, including any access panels;
2 not be of sheet metal;
3 not have any access panels opening into a circulation space or bedroom;
4 be imperforate except for an opening for a pipe or access panel;
5 have the openings for pipes as small as possible and with fire stopping around the pipe.

If the floor being penetrated is both fire resisting and sound insulating, a pipe casing must be provided which embraces both the above sets of requirements.

Chapter 6

ROOFS

| 6.1   The Building Regulations applicable |

The Building Regulations relating to the construction of a roof of a residential building are:

A1   Loading
B2   Internal fire spread (linings)
B3   Internal fire spread (structure)
B4   External fire spread
C4   Resistance to weather and ground moisture
F2   Condensation
H3   Rainwater drainage
L1   Conservation of fuel and power
Reg 7 Materials and workmanship

A1 requires that the building, in this case the roof, be constructed so that the imposed loads are sustained safely; B2 is particularly concerned with the ability of roof lights, plastic or otherwise, to resist the spread of flame over their surface; B3 states that the building must be designed and constructed so that the unseen spread of fire and smoke within concealed spaces in its structure and fabric is inhibited; B4 requires that the roof of a building should resist the spread of fire over the roof and from one building to another; C4 requires that the roof shall resist the passage of moisture to the inside of the building; F2 lays down that adequate provision must be made to prevent condensation in a roof or roof void above a suspended ceiling; H3 is about the adequacy of systems carrying the rainwater from a roof; L1 states that adequate provision must be made for the conservation of fuel and power; and Regulation 7 is concerned with the standard of materials and workmanship employed.

While Regulation H3 relates to roofs, it is also a subject associated with drainage generally and is dealt with in Chapter 13.

Regulation B3 has a significant historical significance since it was the rapidity with which fire spread through the roofs of houses in Pudding Lane that caused the initial ferocity of the Great Fire of London and gave rise to the first legislation on building control as we know it now.

6.2 Weather resistance

The primary purpose of a roof is to exclude the weather. Approved Document C4, Section 5 treats roof finishes as cladding and sets out two requirements. Firstly, that it will resist the penetration of rain and snow to the interior of the building and, secondly, that it will not be damaged by rain or snow nor carry the rain or snow to any part of the building that would be damaged by it.

It is considered that any cladding with overlapping dry joints, i.e. slating and tiling, will meet the performance requirements if it is impervious or weather resisting and backed by a material that will direct any rain or snow that has entered the cladding back towards the outside face. This backing material is, of course. the sarking felt now laid below slating and tiling as standard practice.

Reference is also made in Document C4 to materials which deteriorate rapidly and which should only be used if certain conditions are met. The conditions attached to the use of such materials would be found in the relevant British Standards and Agrément Certificates of Approval set out in the Approved Document supporting Regulation 7, reviewed in Chapter 1.

Where sheet roofing materials are to be used, the requirements can be met by following the recommendations of the relevant British Standard below:

- BS CP 143: *Code of practice for sheet roof and wall coverings*
  Part 1 deals with corrugated and troughed aluminium roofing.
  Part 5 deals with zinc roofing.
  Part 10 deals with galvanized corrugated roofing.
  Part 12 deals with copper roofing.
  Part 15 deals with sheet aluminium roofing.
  Part 16 deals with semi-rigid asbestos bitumen sheet roofing.
- BS 6915: *Specification for design and construction of fully supported lead sheet roof and wall coverings.*

6.3 Structural stability

The load a roof is expected to carry due to snow lying on it varies, naturally, according to the area of the country in which it is located and the altitude of the site. The map in Fig. 6.1 shows those parts of the country and altitudes where the imposed load can be taken as $0.75 \, \text{kN/m}^2$ and those where it must be increased to $1.00 \, \text{kN/m}^2$.

Tables 6.1 to 6.14 give the sizes of timbers in pitched roofs and Tables 6.15 to 6.17 give the sizes of timber in flat roofs all at a variety of spacings and covering both the loadings mentioned above. They are based on Tables A3 to A22 of Approved Document A1/2 but not following the same order and, in some cases, the Approved Document tables have been combined into one.

The timber sizes listed are a selection from the Approved Document on the

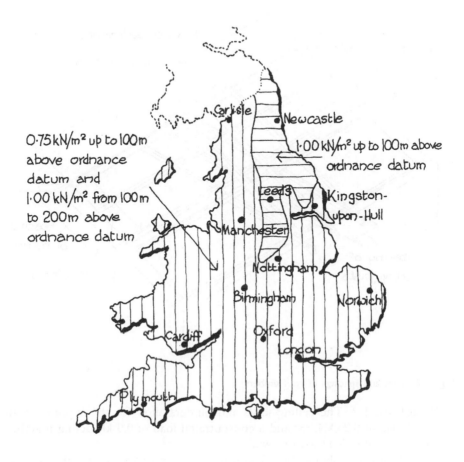

0·75 kN/m² up to 100m above ordnance datum and 1·00 kN/m² from 100m to 200m above ordnance datum

1·00 kN/m² up to 100m above ordnance datum

Fig. 6.1 Map of snow load allowances

basis of those commonly available. The dead loads are also selected from those given in A1/2 to accord with the dead loads normally encountered. For timber sizes or dead loadings not covered, reference should be made to the original relevant table.

The tables do not apply to trussed rafter roofs. For information on this subject see BS 5268; Part 3: *Code of practice for trussed rafter roofs*.

No notches or holes should be cut in any of the pitched roof timbers and only in the flat roof timbers within the limits set out in relation to floor joists in Chapter 5 and illustrated in Fig. 5.2.

Spacing and span of roof timbers is to be taken as shown in Fig. 6.2.

The following notes apply to specific tables as shown:

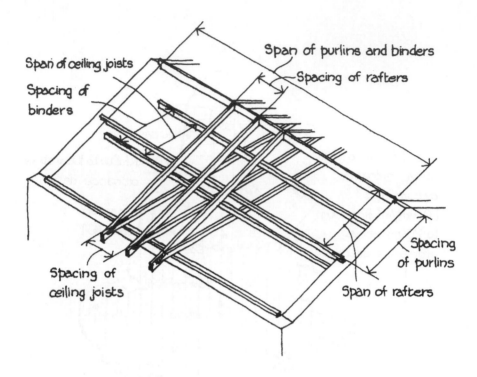

Fig. 6.2 Spans and spacings of roof timbers

Tables 6.1 and 6.2 The ceiling joists and binders allow for the support of an imposed load of 0.25 kN/m^2 and a concentrated load of 0.9 kN acting together in addition to the dead load shown.

Special measures should be taken where it is necessary to trim the timbers for chimney openings and for the support of water tanks. Minimum end bearing for both ceiling joists and binders is 35 mm.

Tables 6.3 to 6.8 The rafters and purlins have been designed to support the imposed load of 0.75 kN/m^2 shown (measured on plan) or a concentrated load of 0.9 kN.

Purlins are to be fixed at right angles to the roof slope. Minimum end bearing for rafters is 35 mm and for purlins is 50 mm.

Tables 6.9 to 6.14 The rafters and purlins have been designed to support the imposed load of 1.00 kN/m^2 shown or a concentrated load of 0.9 kN.

Purlins are to be fixed at right angles to the roof. Minimum end bearing for rafters is 35 mm and for purlins is 50 mm.

Tables 6.15, 6.16 and 6.17 The joists are designed to support the distributed loads shown or a concentrated load of 0.9 kN.

If the wearing surface to be laid on a flat roof to which Table 6.17 relates is of a heavy nature, the total dead load should be carefully checked against the maximum allowed in the table. Minimum end bearing for flat roof joists is 35 mm.

Table 6.1 Spans of ceiling joists

Size of joist	Timber strength class SC3			Timber strength class SC4		
	Maximum clear span in metres			Maximum clear span in metres		
	Spacing in millimetres			Spacing in millimetres		
	400	450	600	400	450	600
38×72	1.15	1.14	1.11	1.21	1.20	1.17
38×97	1.74	1.72	1.67	1.84	1.82	1.76
38×122	2.37	2.34	2.25	2.50	2.46	2.37
38×147	3.02	2.97	2.85	3.18	3.13	3.00
38×170	3.63	3.57	3.41	3.81	3.75	3.58
38×195	4.30	4.23	4.02	4.51	4.43	4.22
38×220	4.98	4.88	4.64	5.21	5.11	4.86
47×72	1.27	1.26	1.23	1.35	1.33	1.30
47×97	1.92	1.90	1.84	2.03	2.00	1.93
47×122	2.60	2.57	2.47	2.74	2.70	2.60
47×147	3.30	3.25	3.11	3.47	3.42	3.27
47×170	3.96	3.89	3.72	4.15	4.08	3.89
47×195	4.68	4.59	4.37	4.90	4.81	4.57
47×220	5.39	5.29	5.03	5.64	5.53	5.25
50×72	1.31	1.30	1.27	1.39	1.37	1.34
50×97	1.97	1.95	1.89	2.08	2.06	1.99
50×122	2.67	2.63	2.53	2.81	2.77	2.66
50×147	3.39	3.34	3.19	3.56	3.50	3.35
50×170	4.06	3.99	3.81	4.25	4.18	3.99
50×195	4.79	4.70	4.48	5.01	4.92	4.68
50×220	5.52	5.41	5.14	5.77	5.66	5.37

(*Source:* Based on Table A3 of Approved Document A1/2)

Table 6.2   Spans of binders supporting ceiling joists

Size of binder (mm)	Timber strength class SC3 — Maximum clear span in metres — Spacing in metres						Timber strength class SC4 — Maximum clear span in metres — Spacing in metres					
	1.2	1.5	1.8	2.1	2.4	2.7	1.2	1.5	1.8	2.1	2.4	2.7
47×150	2.17	2.05	1.96	1.88	1.81		2.28	2.16	2.06	1.98	1.9	1.84
47×175	2.59	2.45	2.33	2.24	2.15	2.08	2.72	2.57	2.45	2.34	2.26	2.18
50×150	2.22	2.11	2.01	1.93	1.86		2.33	2.21	2.11	2.20	1.95	1.89
50×175	2.65	2.51	2.39	2.29	2.21	2.13	2.78	2.63	2.51	2.40	2.31	2.23
50×200	3.08	2.91	2.77	2.65	2.55	2.47	3.23	3.05	2.90	2.78	2.67	2.58
75×125	2.12	2.01	1.92	1.85			2.22	2.11	2.01	1.94	1.87	1.18
75×150	2.61	2.47	2.36	2.26	2.18	2.11	2.73	2.59	2.47	2.37	2.28	2.21
75×175	3.10	2.93	2.79	2.68	2.58	2.49	3.24	3.07	2.92	2.80	2.70	2.61
75×200	3.59	3.39	3.23	3.09	2.98	2.88	3.75	3.54	3.37	3.23	3.11	3.00
75×225	4.08	3.85	3.66	3.51	3.37	3.26	4.26	4.02	3.82	3.66	3.52	3.40

(Source: Based on Table A4 of Approved Document A1/2)

Table 6.3   (Source: Based on Table A5 of Approved Document A1/2)

Pitch 15 to 22.5 degrees	Spans of rafters												Load 0.75 kN/m^2
Size of rafter	Timber strength class SC3 — Maximum clear span in metres for dead loads of:						Timber strength class SC4 — Maximum clear span in metres for dead loads of:						
	under 0.5 kN/m^2 Spacing in millimetres			0.5 to 0.75 kN/m^2 Spacing in millimetres			under 0.5 kN/m^2 Spacing in millimetres			0.5 to 0.75 kN/m^2 Spacing in millimetres		
	400	450	600	400	450	600	400	450	600	400	450	600
38×100	2.10	2.05	1.93	1.93	1.88	1.75	2.42	2.33	2.11	2.28	2.19	1.99
38×125	2.89	2.79	2.53	2.63	2.55	2.34	3.01	2.90	2.64	2.83	2.73	2.48
38×150	3.47	3.34	3.03	3.26	3.14	2.78	3.60	3.47	3.16	3.39	3.26	2.97
47×100	2.46	2.40	2.18	2.25	2.19	2.03	2.59	2.49	2.27	2.44	2.35	2.13
47×125	3.10	2.99	2.72	2.92	2.81	2.56	3.22	3.11	2.83	3.04	2.92	2.66
47×150	3.71	3.57	3.25	3.50	3.36	3.06	3.85	3.71	3.38	3.63	3.50	3.18
50×100	2.54	2.45	2.23	2.35	2.29	2.09	2.64	2.54	2.32	2.49	2.40	2.18
50×125	3.17	3.05	2.78	2.98	2.87	2.61	3.29	3.17	2.89	3.10	2.98	2.72
50×150	3.78	3.64	3.32	3.57	3.43	3.12	3.93	3.78	3.45	3.70	3.57	3.25

Table 6.4   (*Source:* Based on Table A9 of Approved Document A1/2)

Pitch 22.5 to 30 degrees	Spans of rafters											Load 0.75 kN/m²
	Timber strength class SC3						Timber strength class SC4					
	Maximum clear span in metres for dead loads of:						Maximum clear span in metres for dead loads of:					
Size of rafter	under 0.5 kN/m²			0.5 to 0.75 kN/m²			under 0.5 kN/m²			0.5 to 0.75 kN/m²		
	Spacing in millimetres			Spacing in millimetres			Spacing in millimetres			Spacing in millimetres		
	400	450	600	400	450	600	400	450	600	400	450	600
38×100	2.18	2.13	2.01	2.01	1.96	1.82	2.48	2.38	2.17	2.33	2.24	2.03
38×125	2.97	2.86	2.60	2.74	2.66	2.44	3.08	2.97	2.70	2.90	2.79	2.53
38×150	3.55	3.42	3.11	3.34	3.21	2.92	3.69	3.55	3.23	3.47	3.34	3.04
47×100	2.55	2.46	2.23	2.35	2.28	2.10	2.65	2.55	2.32	2.49	2.40	2.18
47×125	3.18	3.06	2.79	2.99	2.88	2.62	3.30	3.18	2.90	3.11	2.99	2.72
47×150	3.80	3.66	3.33	3.57	3.44	3.13	3.94	3.80	3.46	3.71	3.58	3.26
50×100	2.60	2.51	2.28	2.45	2.36	2.14	2.71	2.61	2.37	2.55	2.45	2.23
50×125	3.24	3.12	2.84	3.05	2.93	2.67	3.37	3.24	2.96	3.17	3.05	2.78
50×150	3.87	3.73	3.40	3.65	3.51	3.20	4.02	3.87	3.53	3.79	3.65	3.32

Table 6.5   (*Source:* Based on Table A13 of Approved Document A1/2)

Pitch 30 to 45 degrees	Spans of rafters											Load 0.75 kN/m²
	Timber strength class SC3						Timber strength class SC4					
	Maximum clear span in metres for dead loads of:						Maximum clear span in metres for dead loads of:					
Size of rafter	under 0.5 kN/m²			0.5 to 0.75 kN/m²			under 0.5 kN/m²			0.5 to 0.75 kN/m²		
	Spacing in millimetres			Spacing in millimetres			Spacing in millimetres			Spacing in millimetres		
	400	450	600	400	450	600	400	450	600	400	450	600
38×100	2.28	2.23	2.10	2.10	2.05	1.91	2.56	2.47	2.42	2.40	2.31	2.10
38×125	3.07	2.95	2.69	2.87	2.77	2.52	3.19	3.07	2.80	2.99	2.88	2.62
38×150	3.67	3.53	3.22	3.44	3.31	3.01	3.81	3.67	3.35	3.58	3.45	3.14
47×100	2.64	2.54	2.31	2.45	2.38	2.17	2.74	2.64	2.41	2.58	2.48	2.25
47×125	3.29	3.17	2.88	3.09	2.97	2.70	3.41	3.29	3.00	3.21	3.09	2.81
47×150	3.93	3.78	3.45	3.69	3.55	3.23	4.08	3.93	3.59	3.83	3.69	3.36
50×100	2.69	2.59	2.36	2.53	2.43	2.21	2.80	2.70	2.45	2.63	2.53	2.30
50×125	3.35	3.23	2.94	3.15	3.03	2.76	3.48	3.35	3.06	3.27	3.15	2.87
50×150	4.00	3.86	3.52	3.76	3.62	3.30	4.16	4.01	3.66	3.91	3.77	3.43

Table 6.6   (Source: Based on Table A13 of Approved Document A1/2)

Pitch 15 to 22.5 degrees	Purlins											Load 0.75 kN/m²
	Timber strength class SC3											
Size of purlin	Maximum clear span in metres for dead loads of:											
	not more than 0.5 kN/m²						0.5 to 0.75 kN/m²					
	Spacing of purlin in metres						Spacing of purlin in metres					
	1.50	1.80	2.10	2.40	2.70	3.00	1.50	1.80	2.10	2.40	2.70	3.00
50×150	1.90											
50×175	2.22	2.08	1.96	1.87			2.08	1.95	1.84			
50×200	2.53	2.37	2.24	2.13	2.02	1.92	2.38	2.22	2.10	1.97	1.85	
50×225	2.84	2.66	2.52	2.40	2.26	2.14	2.67	2.50	2.35	2.20	2.07	1.96
63×150	2.06	1.94	1.83				1.94	1.82				
63×175	2.41	2.26	2.13	2.03	1.95	1.87	2.26	2.12	2.00	1.91	1.82	
63×200	2.75	2.58	2.44	2.32	2.22	2.14	2.58	2.42	2.29	2.18	2.08	1.97
63×225	3.09	2.89	2.74	2.61	2.50	2.40	2.90	2.72	2.57	2.45	2.33	2.20
75×125	1.83											
75×150	2.19	2.06	1.95	1.86			2.06	1.94	1.83			
75×175	2.56	2.40	2.27	2.17	2.08	2.00	2.41	2.26	2.13	2.03	1.95	1.87
75×200	2.92	2.74	2.59	2.47	2.37	2.28	2.75	2.58	2.44	2.32	2.22	2.14
75×225	3.28	3.08	2.91	2.78	2.66	2.56	3.09	2.89	2.74	2.61	2.50	2.40
	Timber strength class SC4											
50×150	1.99	1.86					1.87					
50×175	2.32	2.17	2.05	1.95	1.87		2.18	2.04	1.92	1.83		
50×200	2.64	2.48	2.34	2.23	2.14	2.05	2.49	2.33	2.20	2.09	2.00	1.92
50×225	2.97	2.78	2.63	2.51	2.40	2.31	2.79	2.62	2.47	2.35	2.25	2.16
63×150	2.16	2.02	1.91	1.82			2.03	1.90				
63×175	2.51	2.36	2.23	2.13	2.04	1.96	2.36	2.22	2.10	2.00	1.91	1.84
63×200	2.87	2.69	2.55	2.43	2.33	2.24	2.70	2.53	2.39	2.28	2.18	2.10
63×225	3.22	3.02	2.86	2.73	2.61	2.52	3.03	2.84	2.69	2.56	2.45	2.36
75×125	1.91											
75×150	2.29	2.15	1.04	1.94	1.86		2.16	2.02	1.91	1.82		
75×175	2.67	2.51	2.37	2.26	2.17	2.09	2.51	2.36	2.23	2.13	2.04	1.96
75×200	3.05	2.86	2.71	2.58	2.48	2.39	2.87	2.69	2.55	2.43	2.33	2.24
75×225	3.42	3.21	3.04	2.90	2.78	2.68	3.22	3.02	2.86	2.73	2.62	2.52

Table 6.7   (*Source:* Based on Table A10 of Approved Document A1/2)

Pitch 22.5 to 30 degrees	Purlins											Load 0.75 kN/m²

	Timber strength class SC3											
Size of purlin	Maximum clear span in metres for dead loads of:											
	not more than 0.5 kN/m²						0.5 to 0.75 kN/m²					
	Spacing of purlin in metres						Spacing of purlin in metres					
	1.50	1.80	2.10	2.40	2.70	3.00	1.50	1.80	2.10	2.40	2.70	3.00
50×150	1.95	1.83					1.83					
50×175	2.27	2.12	2.01	1.92	1.83		2.13	1.99	1.88			
50×200	2.59	2.43	2.30	2.19	2.09	1.99	2.43	2.28	2.15	2.03	1.91	1.81
50×225	2.92	2.73	2.58	2.46	2.34	2.22	2.74	2.56	2.42	2.27	2.14	2.02
63×150	2.12	1.98	1.88				1.99	1.86				
63×175	2.47	2.31	2.19	2.09	2.00	1.92	2.32	2.17	2.05	1.95	1.87	
63×200	2.81	2.64	2.50	2.38	2.28	2.19	2.64	2.48	2.34	2.23	2.13	2.04
63×225	3.16	2.97	2.81	2.68	2.56	2.47	2.97	2.78	2.63	2.51	2.40	2.28
75×125	1.88											
75×150	2.25	2.11	2.00	1.91	1.83		2.11	1.98	1.87			
75×175	2.62	2.46	2.33	2.22	2.13	2.05	2.46	2.31	2.19	2.08	1.99	1.92
75×200	2.99	2.81	2.66	2.54	2.43	2.34	2.81	2.64	2.50	2.38	2.28	2.19
75×225	3.36	3.15	2.99	2.85	2.73	2.63	3.16	2.96	2.80	2.67	2.56	2.46
	Timber strength class SC4											
50×150	2.04	1.91	1.81				1.91					
50×175	2.37	2.22	2.10	2.00	1.92	1.84	2.23	2.09	1.97	1.88		
50×200	2.71	2.54	2.40	2.29	2.19	2.11	2.54	2.38	2.25	2.14	2.05	1.97
50×225	3.05	2.86	2.70	2.57	2.46	2.37	2.86	2.68	2.53	2.41	2.30	2.21
63×150	2.21	2.07	1.96	1.87			2.08	1.95	1.84			
63×175	2.57	2.42	2.29	2.18	2.09	2.01	2.42	2.27	2.15	2.04	1.96	1.88
63×200	2.94	2.76	2.61	2.49	2.39	2.30	2.76	2.59	2.45	2.33	2.24	2.15
63×225	3.30	3.10	2.93	2.80	2.68	2.58	3.10	2.91	2.75	2.62	2.51	2.42
75×125	1.96	1.84					1.84					
75×150	2.35	2.20	2.09	1.99	1.91	1.84	2.21	2.07	1.96	1.87		
75×175	2.73	2.57	2.43	2.32	2.22	2.14	2.57	2.41	2.28	2.18	2.09	2.01
75×200	3.12	2.93	2.78	2.65	2.54	2.45	2.93	2.75	2.61	2.49	2.38	2.29
75×225	3.50	3.29	3.12	2.98	2.86	2.75	3.30	3.10	2.93	1.80	2.68	2.58

Table 6.8　　　　　　　　(*Source:* Based on Table A14 of Approved Document A1/2)

Pitch 30 to 40 degrees	Purlins											Load 0.75 kN/m²

Timber strength class SC3

Maximum clear span in metres for dead loads of:

Size of purlin	not more than 0.5 kN/m²						0.5 to 0.75 kN/m²					
	Spacing of purlin in metres						Spacing of purlin in metres					
	1.50	1.80	2.10	2.40	2.70	3.00	1.50	1.80	2.10	2.40	2.70	3.00
50×150	2.02	1.89					1.89					
50×175	2.36	2.21	2.09	1.99	1.90	1.83	2.21	2.06	1.95	1.86		
50×200	2.69	2.52	2.38	2.27	2.17	2.09	2.52	2.36	2.23	2.12	2.01	1.90
50×225	3.02	2.83	2.68	2.55	2.44	2.34	2.83	2.65	2.50	2.38	2.24	2.12
63×150	2.19	2.06	1.95	1.85			2.05	1.93	1.82			
63×175	2.55	2.40	2.27	2.16	2.07	1.99	2.39	2.24	2.12	2.02	1.94	1.86
63×200	2.91	2.74	2.59	2.47	2.37	2.28	2.73	2.56	2.42	2.31	2.21	2.13
63×225	3.28	3.07	2.91	2.78	2.66	2.56	3.07	2.88	2.73	2.60	2.49	2.39
75×125	1.94	1.82					1.82					
75×150	2.33	2.19	2.07	1.97	1.89	1.82	2.18	2.05	1.94	1.85		
75×175	2.71	2.55	2.41	2.30	2.21	2.12	2.55	2.39	2.26	2.15	2.06	1.99
75×200	3.10	2.91	2.75	2.63	2.52	2.43	2.91	2.73	2.58	2.46	2.36	2.27
75×225	3.48	3.27	3.10	2.95	2.83	2.73	3.26	3.06	2.90	2.77	2.65	2.55
Timber strength class SC4												
50×150	2.11	1.98	1.87				1.98	1.85				
50×175	2.46	2.31	2.18	2.08	1.99	1.91	2.31	2.16	2.04	1.94	1.86	
50×200	2.81	2.63	2.49	2.37	2.27	2.19	1.63	2.47	2.33	2.22	2.12	2.04
50×225	3.16	2.96	2.80	2.67	2.56	2.46	2.96	2.77	2.62	2.50	2.39	2.30
63×150	2.29	2.15	2.03	1.94	1.86		2.15	2.01	1.90	1.81		
63×175	2.67	2.50	2.37	2.26	2.17	2.08	2.50	2.35	2.22	2.12	2.03	1.95
63×200	3.04	2.86	2.71	2.58	2.47	2.38	2.86	2.68	2.54	2.42	2.31	2.23
63×225	3.42	3.21	3.04	2.90	2.78	2.68	3.21	3.01	2.85	2.72	2.60	2.50
75×125	2.03	1.90	1.80				1.90					
75×150	2.43	2.28	2.16	2.06	1.98	1.91	2.28	2.14	2.03	1.93	1.85	
75×175	2.83	2.66	2.52	2.40	2.31	2.22	2.66	2.49	2.36	2.25	2.16	2.08
75×200	3.23	3.03	2.88	2.74	2.63	2.54	3.03	2.85	2.70	2.57	2.47	2.37
75×225	3.63	3.41	3.23	3.08	2.96	2.85	3.41	3.20	3.03	2.89	2.77	2.67

Table 6.9   (*Source:* Based on Table A7 of Approved Document A1/2)

Pitch 15 to 22.5 degrees	Spans of rafters												Load 1.00 kN/m²

	Timber strength class SC3						Timber strength class SC4					
	Maximum clear span in metres for dead loads of:						Maximum clear span in metres for dead loads of:					
Size of rafter	under 0.5 kN/m²			0.5 to 0.75 kN/m²			under 0.5 kN/m²			0.5 to 0.75 kN/m²		
	Spacing in millimetres			Spacing in millimetres			Spacing in millimetres			Spacing in millimetres		
	400	450	600	400	450	600	400	450	600	400	450	600
38×100	2.10	2.05	1.90	1.93	1.88	1.75	2.28	2.19	1.99	2.16	2.08	1.89
38×125	2.73	2.63	2.35	2.59	2.49	2.17	2.84	2.73	2.48	2.70	2.59	2.35
38×150	3.27	3.14	2.79	3.10	2.97	2.58	3.40	3.27	2.97	3.23	3.10	2.82
47×100	2.35	2.26	2.05	2.23	2.15	1.95	2.44	2.35	2.14	2.32	2.23	2.03
47×125	2.93	2.82	2.56	2.78	2.68	2.41	3.04	2.93	2.67	2.89	2.78	2.53
47×150	3.50	3.37	3.07	3.33	3.20	2.86	3.64	3.50	3.19	3.46	3.33	3.03
50×100	2.40	2.31	2.10	2.28	2.19	1.99	2.49	2.40	2.18	2.37	2.28	2.07
50×125	2.99	2.88	2.62	2.84	2.73	2.48	3.10	2.99	2.72	2.95	2.84	2.58
50×150	3.57	3.44	3.13	3.40	3.27	2.95	3.71	3.57	3.26	3.46	3.40	3.09

Table 6.10   (*Source:* Based on Table A11 of Approved Document A1/2)

Pitch 22.5 to 30 degrees	Spans of rafters												Load 1.00 kN/m²

	Timber strength class SC3						Timber strength class SC4					
	Maximum clear span in metres for dead loads of:						Maximum clear span in metres for dead loads of:					
Size of rafter	under 0.5 kN/m²			0.5 to 0.75 kN/m²			under 0.5 kN/m²			0.5 to 0.75 kN/m²		
	Spacing in millimetres			Spacing in millimetres			Spacing in millimetres			Spacing in millimetres		
	400	450	600	400	450	600	400	450	600	400	450	600
38×100	2.18	2.13	1.96	2.01	1.96	1.82	2.34	2.25	2.04	2.21	2.13	1.93
38×125	2.80	2.69	2.45	2.65	2.55	2.30	2.91	2.80	2.55	2.76	2.66	2.41
38×150	3.35	3.22	2.93	3.18	3.06	2.73	3.48	3.35	3.05	3.30	3.18	2.89
47×100	2.41	2.32	2.11	2.28	2.20	2.00	2.51	2.41	2.19	2.38	2.29	2.08
47×125	3.00	2.89	2.63	2.85	2.74	2.49	3.12	3.00	2.73	2.96	2.85	2.59
47×150	3.59	3.46	3.14	3.41	3.28	2.98	3.37	3.59	3.27	3.54	3.41	3.10
50×100	2.46	2.37	2.15	2.33	2.24	2.04	2.56	2.46	2.24	2.42	2.33	2.12
50×125	3.06	2.95	2.68	2.91	2.80	2.54	3.18	3.06	2.79	3.02	2.91	2.64
50×150	3.66	3.52	3.21	3.48	3.34	3.04	3.80	3.66	3.34	3.61	3.48	3.16

Table 6.11   (Source: Based on Table A15 of Approved Document A1/2)

Pitch 30 to 45 degrees	Spans of rafters												Load 1.00 kN/m²
	Timber strength class SC3						Timber strength class SC4						
	Maximum clear span in metres for dead loads of:						Maximum clear span in metres for dead loads of:						
Size of rafter	under 0.5 kN/m²			0.5 to 0.75 kN/m²			under 0.5 kN/m²			0.5 to 0.75 kN/m²			
	Spacing in millimetres			Spacing in millimetres			Spacing in millimetres			Spacing in millimetres			
	400	450	600	400	450	600	400	450	600	400	450	600	
38×100	2.28	2.23	2.03	2.10	2.05	1.91	2.42	2.33	2.12	2.29	2.20	2.00	
38×125	2.90	2.79	2.54	2.75	2.64	2.40	3.02	2.90	2.64	2.86	2.75	2.50	
38×150	3.47	3.34	3.04	3.29	3.16	2.87	3.61	3.47	3.16	3.42	3.29	2.99	
47×100	2.50	2.40	2.18	2.36	2.27	2.06	2.60	2.50	2.27	2.46	2.36	2.15	
47×125	3.11	2.99	2.72	2.94	2.83	2.58	3.23	3.11	2.83	3.06	2.95	2.68	
47×150	3.72	3.58	3.26	3.52	3.39	3.08	3.86	3.72	3.39	3.66	3.52	3.21	
50×100	2.55	2.45	2.23	2.41	2.32	2.11	2.65	2.55	2.32	2.51	2.41	2.19	
50×125	3.17	3.05	2.78	3.00	2.89	2.63	3.30	3.17	2.89	3.12	3.01	2.73	
50×150	3.79	3.65	3.33	3.59	3.46	3.15	3.94	3.79	3.46	3.73	3.60	3.27	

Table 6.12   (*Source:* Based on Table A8 of Approved Document A1/2)

Pitch 15 to 22.5 degrees	Purlins											Load 1.00 kN/m²
	Timber strength class SC3											
Size of purlin	*Maximum clear span in metres for dead loads of:*											
	not more than 0.5 kN/m²						*0.5 to 0.75 kN/m²*					
	Spacing of purlin in metres						*Spacing of purlin in metres*					
	1.50	*1.80*	*2.10*	*2.40*	*2.70*	*3.00*	*1.50*	*1.80*	*2.10*	*2.40*	*2.70*	*3.00*
50×175	2.09	1.95	1.84				1.97	1.85				
50×200	2.38	2.23	2.10	1.97	1.85		2.26	2.11	1.96	1.82		
50×225	2.68	2.50	2.36	2.20	2.07	1.96	2.54	2.36	2.18	2.04	1.92	1.81
63×150	1.94	1.82					1.84					
63×175	2.27	2.12	2.01	1.91	1.83		2.15	2.01	1.90	1.81		
63×200	2.59	2.42	2.29	2.18	2.09	1.98	2.45	2.30	2.17	2.06	1.94	1.83
63×225	2.91	2.72	2.58	2.45	2.33	2.21	2.76	2.58	2.44	2.30	2.16	2.05
75×150	2.07	1.94	1.83				1.96	1.84				
75×175	2.41	2.26	2.14	2.04	1.95	1.88	2.29	2.14	2.03	1.93	1.85	
75×200	2.75	2.58	2.44	2.33	2.23	2.14	2.61	2.45	2.31	2.20	2.11	2.01
75×225	3.09	2.90	2.74	2.61	2.50	2.41	2.93	2.75	2.60	2.48	2.36	2.24
	Timber strength class SC4											
50×150	1.87											
50×175	2.18	2.04	1.93	1.83			2.07	1.93	1.82			
50×200	2.49	2.33	2.20	2.10	2.00	1.92	2.36	2.21	2.08	1.98	1.89	
50×225	2.80	2.62	2.48	2.36	2.25	2.16	2.65	2.48	2.34	2.23	2.13	1.95
63×150	2.03	1.90	1.80				1.93	1.81				
63×175	2.37	2.22	2.10	2.00	1.91	1.84	2.25	2.10	1.99	1.89	1.81	
63×200	2.70	2.53	2.40	2.28	2.19	2.10	2.57	2.40	2.27	2.16	2.07	1.99
63×225	3.04	2.85	2.70	2.57	2.46	2.36	2.88	2.70	2.55	2.43	2.32	2.23
75×125	1.80											
75×150	2.16	2.03	1.92	1.83			2.05	1.92	1.82			
75×175	2.52	2.36	2.24	2.13	2.04	1.96	2.39	2.24	2.12	2.02	1.93	1.86
75×200	2.87	2.70	2.55	2.43	2.33	2.24	2.73	2.56	2.42	2.31	2.21	2.12
75×225	3.23	3.03	2.87	2.74	2.62	2.52	3.07	2.88	2.72	2.59	2.48	2.39

Table 6.13   (*Source:* Based on Table A12 of Approved Document A1/2)

Pitch 22.5 to 30 degrees	Purlins											Load 1.00 kN/m²
	Timber strength class SC3											
Size of purlin	*Maximum clear span in metres for dead loads of:*											
	not more than 0.5 kN/m²						*0.5 to 0.75 kN/m²*					
	Spacing of purlin in metres						*Spacing of purlin in metres*					
	1.50	*1.80*	*2.10*	*2.40*	*2.70*	*3.00*	*1.50*	*1.80*	*2.10*	*2.40*	*2.70*	*3.00*
50×175	2.14	2.00	1.89				2.03	1.89				
50×200	2.45	2.29	2.16	2.05	1.93	1.82	2.31	2.16	2.03	1.89		
50×225	2.75	2.57	2.43	2.29	2.15	2.04	2.60	2.43	2.26	2.11	1.99	1.88
63×150	2.00	1.87					1.89					
63×175	2.33	2.18	2.06	1.96	1.88	1.80	2.20	2.06	1.95	1.85		
63×200	2.66	2.49	2.35	2.24	2.14	2.05	2.51	2.35	2.22	2.12	2.00	1.90
63×225	2.98	2.80	2.65	2.52	2.41	2.29	2.83	2.65	2.50	2.38	2.24	2.12
75×150	2.12	1.99	1.88				2.01	1.88				
75×175	2.47	2.32	2.20	2.09	2.00	1.93	2.34	2.20	2.08	1.98	1.89	1.82
75×200	2.82	2.65	2.51	2.39	2.29	2.20	2.68	2.51	2.37	2.26	2.16	2.08
75×225	3.17	2.98	2.82	2.68	2.57	2.47	3.01	2.82	2.67	2.54	2.43	2.32
	Timber strength class SC4											
50×150	1.92						1.82					
50×175	2.24	2.10	1.98	1.89	1.80		2.12	1.98	1.87			
50×200	2.56	2.39	2.26	2.15	2.06	1.98	2.42	2.26	2.14	2.03	1.94	1.86
50×225	2.87	2.69	2.54	2.42	2.32	2.22	2.72	2.55	2.40	2.29	2.18	2.09
63×150	2.09	1.59	1.85				1.98	1.85				
63×175	2.43	2.28	2.16	2.05	1.97	1.89	2.30	2.16	2.04	1.94	1.86	
63×200	2.77	2.60	2.46	2.35	2.25	2.16	2.63	2.46	2.33	2.22	2.12	2.04
63×225	3.12	2.92	2.77	2.64	2.52	2.43	2.95	2.77	2.62	2.49	2.39	2.29
75×150	2.22	2.08	1.97	1.88			2.10	1.97	1.86			
75×175	2.58	2.42	2.29	2.19	2.10	2.02	2.45	2.30	2.17	2.07	1.98	1.91
75×200	2.95	2.77	2.62	2.50	2.39	2.30	2.80	2.62	2.48	2.36	2.26	2.18
75×225	3.31	3.11	2.94	2.81	2.70	2.59	3.14	2.95	2.79	2.66	2.55	2.45

Table 6.14   (*Source*: Based on Table A16 of Approved Document A1/2)

Pitch 30 to 45 degrees	Purlins											Load 1.00 kN/m²

Timber strength class SC3

	Maximum clear span in metres for dead loads of:											
Size of purlin	not more than 0.5 kN/m²						0.5 to 0.75 kN/m²					
	Spacing of purlin in metres						Spacing of purlin in metres					
	1.50	1.80	2.10	2.40	2.70	3.00	1.50	1.80	2.10	2.40	2.70	3.00
50×175	2.22	2.08	1.97				2.10	1.96	1.85			
50×200	2.54	2.38	2.25	2.14	2.03	1.95	2.40	2.24	2.12	1.99	1.87	
50×225	2.85	2.67	2.53	2.40	2.27	2.15	2.70	2.52	2.38	2.22	2.09	1.98
63×150	2.07	1.94	1.84				1.96	1.83				
63×175	2.41	2.26	2.14	2.04	1.95	1.88	2.28	2.14	2.02	1.92	1.84	
63×200	2.76	2.58	2.44	2.33	2.23	2.14	2.61	2.44	2.31	2.20	2.10	2.00
63×225	3.10	2.90	2.75	2.62	2.51	2.41	2.93	2.74	2.59	2.47	2.36	2.23
75×150	2.20	2.07	1.96	1.86			2.08	1.95	1.85			
75×175	2.57	2.41	2.28	2.17	2.08	2.00	2.43	2.28	2.15	2.05	1.96	1.89
75×200	2.93	2.75	2.60	2.48	2.38	2.29	2.77	2.60	2.46	2.34	2.24	2.16
75×225	3.29	3.09	2.92	2.79	2.67	2.57	3.12	2.92	2.76	2.63	2.52	2.43
Timber strength class SC4												
50×150	1.99	1.87					1.88					
50×175	2.32	2.18	2.06	1.96	1.88	1.80	2.20	2.06	1.94	1.85		
50×200	2.65	2.49	2.35	2.24	2.14	2.06	2.51	2.35	2.22	2.11	2.02	1.94
50×225	2.98	2.79	2.64	2.52	2.41	2.31	2.82	2.64	2.49	2.37	2.27	2.18
63×150	2.16	2.03	1.92	1.83			2.05	1.92	1.81			
63×175	2.52	2.36	2.24	2.13	2.04	1.97	2.39	2.24	2.11	2.01	1.93	1.85
63×200	2.88	2.70	2.56	2.44	2.33	2.24	2.72	2.55	2.41	2.30	2.20	2.12
63×225	3.23	3.03	2.87	2.74	2.62	2.52	3.06	2.87	2.71	2.59	2.48	2.38
75×150	2.30	2.16	2.04	1.95	1.87		2.18	2.04	1.93	1.84		
75×175	2.68	2.51	2.38	2.27	2.18	2.10	2.54	2.38	2.25	2.15	2.06	1.98
75×200	3.06	2.87	2.72	2.59	2.49	2.39	2.89	2.72	2.57	2.45	2.35	2.26

Table 6.15   (Source: Based on Tables A17 and A18 of Approved Document A1/2)

Access only for maintenance and repair	Flat roof joists											Loading 0.75 kN/m^2
	Timber strength class SC3						Timber strength class SC4					
	Maximum clear span in metres for dead loads of:						Maximum clear span in metres for dead loads of:					
Size of joist	under 0.5 kN/m^2			0.5 to 0.75 kN/m^2			under 0.5 kN/m^2			0.5 to 0.75 kN/m^2		
	Spacing in millimetres			Spacing in millimetres			Spacing in millimetres			Spacing in millimetres		
	400	450	600	400	450	600	400	450	600	400	450	600
38×97	1.74	1.72	1.67	1.67	1.64	1.58	1.84	1.82	1.76	1.76	1.73	1.66
38×122	2.37	2.34	2.25	2.25	2.21	2.11	2.50	2.46	2.37	2.37	2.33	2.22
38×147	3.02	2.97	2.85	2.85	2.80	2.66	3.18	3.13	3.00	3.00	2.94	2.79
38×170	3.63	3.57	3.37	3.41	3.34	3.17	3.81	3.75	3.50	3.58	3.51	3.30
38×195	4.30	4.23	3.86	4.03	3.94	3.63	4.51	4.40	4.01	4.22	4.13	3.78
38×220	4.94	4.76	4.34	4.64	4.49	4.09	5.13	4.95	4.51	4.85	4.67	4.25
47×97	1.92	1.90	1.84	1.84	1.81	1.74	2.03	2.00	1.94	1.94	1.91	1.83
47×122	2.60	2.57	2.47	2.47	2.43	2.31	2.74	2.70	2.60	2.60	2.55	2.43
47×147	3.30	3.25	3.12	3.12	3.06	2.90	3.47	3.42	3.26	3.27	3.21	3.04
47×170	3.96	2.89	3.61	3.72	3.64	3.40	4.15	4.08	3.76	3.89	3.81	3.54
47×195	4.68	4.53	4.13	4.37	4.28	3.89	4.88	4.70	4.29	4.58	4.44	4.05
47×220	5.28	5.09	4.65	4.99	4.81	4.38	5.48	5.29	4.83	5.18	5.00	4.56
50×97	1.97	1.95	1.89	1.89	1.86	1.78	2.08	2.06	1.99	1.99	1.96	1.88
50×122	2.67	2.64	2.53	2.53	2.49	2.37	2.81	2.77	2.66	2.66	2.62	2.49
50×147	3.39	3.34	3.19	3.19	3.13	2.97	3.56	3.50	3.32	3.35	3.29	3.12
50×170	4.06	3.99	3.69	3.81	3.73	3.47	4.26	4.18	3.83	3.99	3.91	3.61
50×195	4.79	4.62	4.22	4.48	4.36	3.97	4.97	4.80	4.38	4.68	4.53	4.13
50×220	5.38	5.19	4.74	5.09	4.90	4.47	5.59	5.39	4.93	5.28	5.09	4.65
75×122	3.17	3.12	3.00	3.00	2.94	2.80	3.33	3.27	3.14	3.14	3.08	2.93
75×147	3.98	3.92	3.64	3.75	3.67	3.44	4.17	4.10	3.78	3.92	3.84	3.57
75×170	4.74	4.58	4.19	4.44	4.33	3.96	4.92	4.75	4.35	4.64	4.50	4.11
75×195	5.42	5.23	4.79	5.13	4.95	4.53	5.61	5.42	4.97	5.32	5.14	4.70
75×220	6.07	5.87	5.38	5.76	5.56	5.09	6.29	6.08	5.59	5.97	5.77	5.28

Table 6.16   (Source: Based on Tables A19 and A20 of Approved Document A1/2)

Access only for maintenance and repair	Flat roof joists											Loading 1.00 kN/m²

	Timber strength class SC3						Timber strength class SC4					
	Maximum clear span in metres for dead loads of:						Maximum clear span in metres for dead loads of:					
Size of joist	under 0.5 kN/m²			0.5 to 0.75 kN/m²			under 0.5 kN/m²			0.5 to 0.75 kN/m²		
	Spacing in millimetres			Spacing in millimetres			Spacing in millimetres			Spacing in millimetres		
	400	450	600	400	450	600	400	450	600	400	450	600
38×97	1.74	1.72	1.67	1.67	1.64	1.58	1.84	1.82	1.76	1.76	1.73	1.66
38×122	2.37	2.34	2.25	2.25	2.21	2.11	2.50	2.46	2.37	2.37	2.33	2.22
38×147	3.02	2.97	2.75	2.85	2.80	2.61	3.18	3.13	2.86	3.00	2.94	2.71
38×170	3.62	3.49	3.17	3.41	3.31	3.01	3.77	3.63	3.30	3.58	3.45	3.13
38×195	4.15	3.99	3.63	3.94	3.79	3.45	4.31	4.15	3.78	4.10	3.95	3.59
38×220	4.67	4.49	4.09	4.44	4.27	3.88	4.85	4.67	4.25	4.61	4.44	4.04
47×97	1.92	1.90	1.84	1.84	1.81	1.74	2.03	2.00	1.94	1.94	1.91	1.83
47×122	2.60	2.57	2.45	2.47	2.43	2.31	2.74	2.70	2.55	2.60	2.55	2.42
47×147	3.30	3.24	2.95	3.12	3.06	2.80	3.47	3.37	3.07	3.27	3.21	2.91
47×170	3.88	3.74	3.40	3.69	3.56	3.23	4.03	3.88	3.54	3.84	3.70	3.36
47×195	4.44	4.27	3.89	4.23	4.07	3.70	4.61	4.44	4.05	4.39	4.23	3.85
47×220	4.99	4.81	4.38	4.75	4.58	4.17	5.18	5.00	4.56	4.94	4.76	4.33
50×97	1.97	1.95	1.89	1.89	1.86	1.78	2.08	2.06	1.99	1.99	1.96	1.88
50×122	2.67	2.64	2.50	2.53	2.49	2.37	2.81	2.77	2.60	2.66	2.62	2.47
50×147	3.39	3.31	3.01	3.19	3.13	2.86	3.56	3.44	3.13	3.35	3.27	2.97
50×170	3.96	3.81	3.47	3.77	3.63	3.30	4.11	3.96	3.61	3.92	3.77	3.43
50×195	4.53	4.36	3.97	4.31	4.15	3.78	4.70	4.53	4.13	4.48	4.31	3.93
50×220	5.09	4.90	4.47	4.85	4.67	4.25	5.28	5.09	4.65	5.04	4.85	4.42
75×122	3.17	3.12	2.86	3.00	2.94	2.72	3.33	3.26	2.97	3.14	3.08	2.83
75×147	3.90	3.76	3.44	3.72	3.59	3.27	4.05	3.91	3.57	3.86	3.72	3.40
75×170	4.49	4.33	3.96	4.29	4.13	3.77	4.66	4.50	4.11	4.45	4.29	3.92
75×195	5.13	4.95	4.53	4.89	4.72	4.31	5.32	5.14	4.70	5.08	4.90	4.48
75×220	5.76	5.56	5.09	5.50	5.30	4.85	5.97	5.77	5.28	5.70	5.50	5.04

Table 6.17   (Source: Based on Tables A21 and A22 of Approved Document A1/2)

Access only for maintenance and repair	Flat roof joists												Loading 1.50 kN/m²
	Timber strength class SC3						Timber strength class SC4						
Size of joist	Maximum clear span in metres for dead loads of:						Maximum clear span in metres for dead loads of:						
	under 0.5 kN/m²			0.5 to 0.75 kN/m²			under 0.5 kN/m²			0.5 to 0.75 kN/m²			
	Spacing in millimetres			Spacing in millimetres			Spacing in millimetres			Spacing in millimetres			
	400	450	600	400	450	600	400	450	600	400	450	600	
38×122	1.74	1.71	1.65	1.68	1.65	1.57	1.86	1.84	1.79	1.81	1.79	1.73	
38×147	2.27	2.25	2.18	2.21	2.18	2.09	2.40	2.38	2.30	2.33	2.30	2.21	
38×170	2.77	2.74	2.64	2.68	2.64	2.53	2.93	2.89	2.79	2.83	2.79	2.67	
38×195	3.33	3.28	3.16	3.21	3.16	3.02	3.51	3.46	3.29	3.38	3.33	3.18	
38×220	3.90	3.84	3.56	3.75	3.68	3.43	4.10	4.04	3.71	3.94	3.87	3.58	
47×122	1.94	1.93	1.87	1.89	1.87	1.81	2.06	2.04	1.98	2.00	1.98	1.91	
47×147	2.51	2.48	2.40	2.44	2.40	2.31	2.66	2.62	2.54	2.57	2.54	2.44	
47×170	3.06	3.02	2.91	2.95	2.91	2.78	3.22	3.18	3.06	3.11	3.06	2.93	
47×195	3.66	3.61	3.40	3.52	3.46	3.28	3.85	3.80	3.54	3.71	3.64	3.42	
47×220	4.27	4.20	3.83	4.10	4.03	3.70	4.49	4.39	3.99	4.31	4.23	3.85	
50×122	2.00	1.98	1.93	1.95	1.93	1.86	2.12	2.10	2.04	2.06	2.04	1.97	
50×147	2.59	2.56	2.47	2.51	2.47	2.38	2.73	2.70	2.61	2.65	2.61	2.51	
50×170	3.14	3.10	2.99	3.04	2.99	2.86	3.31	3.27	3.15	3.20	3.15	3.01	
50×195	3.76	3.70	3.47	3.62	3.56	3.35	3.96	3.90	3.61	3.81	3.74	3.49	
50×220	4.38	4.30	3.91	4.21	4.13	3.78	4.61	4.47	4.07	4.42	4.32	3.93	
75×122	2.42	2.40	3.32	2.35	2.32	2.24	2.56	2.53	2.45	2.48	2.45	2.36	
75×147	3.11	3.07	2.96	3.00	2.96	2.84	3.27	3.23	3.11	3.16	3.11	2.98	
75×170	3.75	3.69	3.47	3.61	3.55	3.35	3.94	3.88	3.61	3.79	3.73	3.49	
75×195	4.45	4.36	3.97	4.28	4.20	3.84	4.67	4.53	4.13	4.49	4.38	3.99	
75×220	5.09	4.90	4.47	4.92	4.74	4.32	5.28	5.09	4.65	5.11	4.93	4.49	

6.4   Thermal insulation

The insulation of the roof of a building contributes a great deal to the capacity of the fabric to conserve fuel and power as required by Regulation L. For this reason, the insulation standard for this particular element in the structure is higher than for any other. The standard to be achieved is expressed as a U value, which is a measure of the rate of heat loss through the structure (see Chapter 3 for an explanation of U value).

All houses and flats require to have an Energy Rating as explained in Chapter 1 Sections 1.9 and 1.10 and that Rating determines the standard U value for the roof. In a building with a pitched roof and a Rating of 60 or less, the U value must be not more than 0.2 W/m²K; if it is over 60, the maximum U value should be 0.25 W/m²K. There is a slight relaxation of this in that the sloping parts of a room-in-the-roof construction may have a value of not more than 0.35 W/m²K.

A flat roof has the same standard as the room-in-the-roof, i.e. 0.35 W/m²K.

Roofs with higher or lower U values can be provided but if it is higher, (i.e. more heat can escape) it is necessary to show that there is a corresponding compensatory better standard of insulation elsewhere in the building. This can be done by the use of the Target U value Method (see Section 1.10.2). Alternatively, a lower value could be used to offset a higher value in another element of the structure.

Table 6.18 shows the commonly available insulating materials and the thicknesses required in timber framed pitched and flat roofs. The Tables in the Approved Document give additional design U values and insulation thicknesses for flat concrete roofs.

6.4.1 Condensation in pitched roofs

Approved Document F2 gives guidance on minimizing condensation in roofs pitched at 15° or more and refers to further detailed guidance in the BRE Report: *Thermal insulation: avoiding the risks*. This gives the following advice:

Condensation is formed when warm, moist air meets a surface or point within a material where the temperature is low. This can often be seen as water running down a window pane. If it is allowed to happen in a roof the resulting moisture can cause serious damage to the members of the structure.

There are two ways to reduce the dangers. Either the moisture laden air must be prevented from reaching the colder parts by the use of vapour check membranes placed on the warm side of any insulation or to take the damp air away before it can cause trouble by ventilating the spaces where it can collect. The former method is used where there are rooms in the roof and the latter is generally used when there is just a roof space.

The BRE Report recommends that ventilation of a roof should be provided by openings in the eaves' soffit which are equivalent to a continuous gap of 10 mm wide. Each opening should be fitted with a 3 to 4 mm mesh to prevent the entry of insects. Care should be taken to ensure that the airway from the eaves into the roof is not blocked by sagging sarking felt or thermal insulating materials and that the air is not polluted by a flue outlet placed directly beneath the eaves' aperture.

It is also recommended that, to minimize the problem, all holes in the ceiling should be blocked so that there is no easy route for the moisture laden air into the roof space. This involves carefully stopping up around all water pipes and electrical cables through the ceiling, particularly in the bathroom, and fitting the loft hatch with a draughtseal which is efficiently compressed by bolts or catches.

There is also the possibility that a cold bridge can be formed at the eaves if the roof insulation does not meet the wall insulation or cavity closure and at the gables where there can be a small gap between the last ceiling joist and the wall.

Figure 6.3 shows details of the recommended constructions.

6.4.2 Condensation in flat roofs

Three forms of flat roof are detailed in the Report:

- Cold deck
- Warm deck (sandwich)
- Warm deck (inverted)

A cold deck roof has the insulation placed between the ceiling and the roof deck, a sandwiched warm deck roof has the insulation laid on a vapour barrier on the roof deck and is covered with the weather-proofing finish and an inverted warm deck roof has the insulation placed on top of the weather-proof layer. Figure 6.4 shows the three typical versions of these forms of construction.

The cold deck system is regarded as the least desirable as there is a high and unpredictable risk of condensation due to the possibility of poor ventilation in windless conditions, even if the roof is correctly designed and constructed. The vapour check should be sealed to the walls around the edges and wherever any services pass through it. It is better to avoid making any holes in the vapour check by arranging to run all services below it. Cross ventilation openings, equivalent to a continuous 25 mm gap, must be provided in opposite eaves between every pair of joists, connecting to an unobstructed 50 mm air space. Any form of strutting which will reduce this airway must be avoided and if a cavity barrier is required for fire resisting purposes, a cold deck roof system should not be used. If the roof form is complex or the roof is on a low building in a sheltered location the ventilation openings should be increased to 0.6 per cent of the roof plan area and, in all cases, an insect proof mesh should be provided. In all flat roofs there is a chance that a cold bridge can occur at the perimeter between the wall insulation and that of the roof leading to severe condensation in the upper angles of the room. To avoid this, care should be taken to see that the roof insulation overlaps whatever insulation is provided in the wall (without blocking either the ventilation openings or the airways leading from them).

The warm deck, sandwich type of roof is not so liable to condensation problems but placing the weather-proof membrane directly on the insulation can cause greater fluctuation in the surface temperature because on a sunny day the insulation prevents the roofing from losing its heat to the room and on

Table 6.18 Thickness of insulation (mm) required in timber pitched and flat roofs

Insulation material	Phenolic foam board		Polyurethane board		Expanded polystyrene slab, Glass fibre slab, Mineral fibre slab		Glass fibre quilt	
Thermal conductivity	0.020 W/mK		0.025 W/mK		0.035 W/mK		0.040 W/mK	
Location of the insulation	Between joists or rafters	Between and over joists or rafters	Between joists or rafters	Between and over joists or rafters	Between joists or rafters	Between and over joists or rafters	Between joists or rafters	Between and over joists or rafters
Basic thickness of insulation for a design U value of:								
0.20 W/m²K	167	126	209	145	293	187	335	209
0.25 W/m²K	114	106	142	120	199	152	227	169
0.35 W/m²K	69	69	86	86	120	112	137	124
Allowable reductions for pitched roof components								
Roof space	4		5		6		7	
19 mm roof tiles	0		1		1		1	
13 mm sarking board	2		2		3		4	
12 mm calcium silicate liner board	1		2		2		3	
10 mm plasterboard	1		2		2		3	
13 mm plasterboard	2		2		3		3	
Allowable reductions for timber flat roof components								
Roof space	3		4		6		6	
19 mm timber deck	3		3		5		5	
19 mm asphalt	1		1		1		2	
3 layer built-up felt	1		1		1		2	
10 mm plasterboard	1		2		2		3	
13 mm plasterboard	2		2		3		3	

(*Source:* Based on Tables A1, A2, A3 and A4 of Approved Document L: 1995 Edition)

Notes: The thicknesses given are based on a roof frame composed of 48 mm wide timbers at 600 mm centres. For other widths and spacing of timber, the U value can be calculated by reference to the procedure given in Appendix B of the Approved Document.

a cold night it stops it gaining any warmth from below. For this reason a high tensile membrane to BS 747, Type 5 should be used if the finish is built-up felt roofing and the first layer should be partially bonded. An asphalt roof finish should be laid on sheathing felt and the insulant must be either selected for its resistance to the heat of the asphalt being laid or protected with a layer of bitumen impregnated fibreboard, corkboard or vermiculite board. As with the

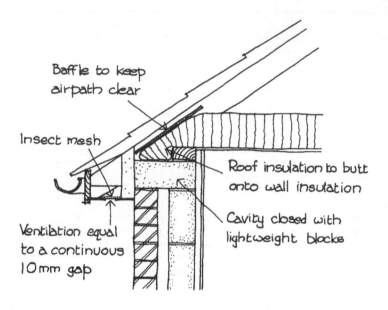

Baffle to keep
airpath clear

Insect mesh

Ventilation equal
to a continuous
10mm gap

Roof insulation to butt
onto wall insulation

Cavity closed with
lightweight blocks

Fig. 6.3 Ventilation of pitched roofs

cold deck roof, provision must be made to connect the wall and roof insulation at the perimeter of the roof to prevent a cold bridge.

As the insulating material in an inverted warm deck roof is above the weather-proof membrane it is subject to a freezing/thawing cycle, wetting and the effects of ultraviolet light. Careful selection of the insulant is, therefore, essential, particularly with respect to water absorption and resistance to frost – the UV light problem is usually solved by the layer of ballast, which is required to prevent the insulation from blowing away, also preventing the light from reaching the insulation. The thickness of the insulation should be increased by 20 per cent to offset the cooling effect of rain running over the surface, particularly in winter. There is also a risk of localized cooling and condensation on the underside of lightweight roof decks if rain or snow can percolate down to the weather-proof layer. This is not usually a risk with concrete roofs because of their greater mass but with timber or metal decked roofs it is recommended that the construction beneath the insulation possesses a thermal resistance of at least $0.15 \, m^2K/W$ and all insulation boards are tightly butted to each other and at projections or upstands. The geotextile membrane shown in Fig. 6.4 is required to contain the ballast and to prevent it from washing down between the insulating boards where it could, eventually, puncture the weather-proofing. This membrane must be turned up at all edges and abutments.

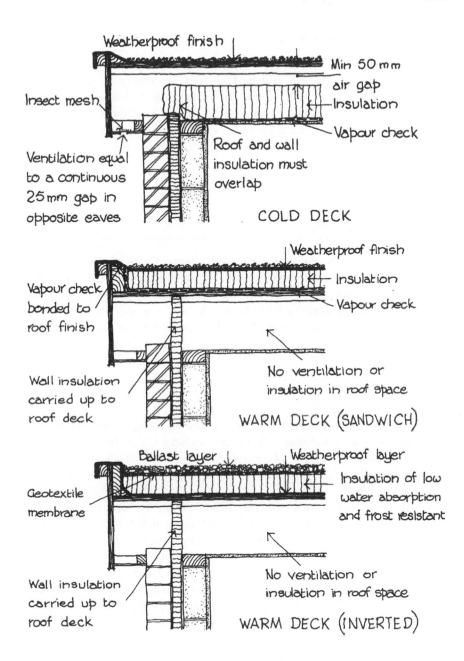

Weatherproof finish

Min 50 mm air gap

Insulation

Vapour check

Insect mesh

Ventilation equal to a continuous 25 mm gap in opposite eaves

Roof and wall insulation must overlap

COLD DECK

Weatherproof finish

Insulation

Vapour check

Vapour check bonded to roof finish

Wall insulation carried up to roof deck

No ventilation or insulation in roof space

WARM DECK (SANDWICH)

Ballast layer

Weatherproof layer

Insulation of low water absorption and frost resistant

Geotextile membrane

Wall insulation carried up to roof deck

No ventilation or insulation in roof space

WARM DECK (INVERTED)

Fig. 6.4 Flat roof constructions

6.4.3 Sealing loft hatches

Leakage of air from the heated parts of the building to unheated roof voids is a source of energy loss which Approved Document L states should be limited. To achieve this with what is the main point of leakage, the loft hatch, it is necessary to fit draught sealing strip. This, it is recommended, should be attached to the stop on which the hatch door sits and a catch or bolt should be fitted in such a way that the sealing strip is compressed when the hatch is closed.

6.5   Fire resistance

Considerations of the dangers of fire with respect to a roof take three main forms: the spread of flame over the outside surface allowing the fire to carry to other buildings, fire within the roof spaces and cavities allowing it to spread to other rooms or buildings within the block; and fire spreading across ceiling and roof lights from one part of the building or room (particularly escape routes – see Chapter 15).

6.5.1   Roof coverings

The principal concerns are the effect on the coverings when exposed to fire from the outside and distance from the boundary. The boundary formed by the wall separating a pair of semi-detached houses is disregarded for this purpose but specific requirements apply to the roof construction at this point (see Section 6.5.2).

There are no limitations as to the distance from the boundary of any of the traditional types of domestic pitched roof constructed with timber framing, covered with sarking felt with or without any type of boarding and finished with slates, clay tiles, concrete tiles or fibre cement slates. Nor are there any similar limitations if the roof is of timber, covered with tongued and grooved boarding and finished with any of the metal roofing systems. If, however, the building has a pitched roof covered with built up bituminous felt roofing there may be limits as to how close to the boundary it can be placed depending on the type of felt to be used and the type of deck on which it is to be laid. Any type of felt upper or lower layers laid over a deck of compressed straw slabs or screeded woodwool slabs is acceptable at any distance from the boundary but if the deck is of 6 mm plywood, 12.5 mm chipboard, 16 mm tongued and grooved timber boarding or 19 mm plain edged timber boarding the distance to the boundary must be not less than:

Type of upper layer	Type of under layer	Minimum distance from the boundary
Type 1E	Type 1B minimum mass 13 kg/10 m^2	6 m unless it is a house in a terrace of three or more in which case 20 m applies
Type 2E	Type 1B minimum mass 13 kg/10 m^2	6 m
Type 2E	Type 2B	No limit
Type 3E	Type 3B or 3G	6 m

The types of roofing felt referred to above are described in BS 747.

Flat roofs with a built up felt roofing finish laid over any type of deck can be sited as near to the boundary as required provided that the felt is covered with 12.5 mm of bitumen bedded stone chippings, bitumen bedded tiles of a non-combustible nature, a sand/cement screed or macadam.

6.5.2 Roof structure

The main area of legislation related to the fire resistance of roof structures concerns the party wall between semi-detached or terraced houses (see Fig. 6.5).

The intention is to prevent a fire in one house running through the roofs of adjoining properties, and to achieve this, the party wall is to be considered as a compartment wall (see Chapter 7) and carried right up to the underside of the roof covering or deck with fire stopping where necessary to maintain the continuity of the fire resistance. Furthermore, the roof covering itself within a zone 1.5 m wide on each side of the wall must possess a high degree of fire resistance.

Natural slates, fibre cement slates and clay or concrete tiles are all of an acceptable fire resisting standard to be laid over a party wall. So is a built-up felt roof covering on either a pitched or flat roof provided that it is laid on compressed straw slabs or screeded woodwool slabs. If the deck is 6 mm plywood, 12.5 mm chipboard, 16 mm tongued and grooved boarding or 19 mm square edged boarding the only acceptable felt roof specification for this purpose is an under layer of type 2E felt and upper layer(s) of type 2B.

The fire stopping can consist of cement mortar; gypsum based plaster; cement or gypsum based vermiculite mixes; glass fibre, crushed rock, blast furnace slag or ceramic based products – with or without resin binders and any of the range of intumescent mastic fillers.

As an alternative to the constructions set out above, the party wall may be extended up through the roof for a height of at least 375 mm above the top surface of the adjoining roof covering.

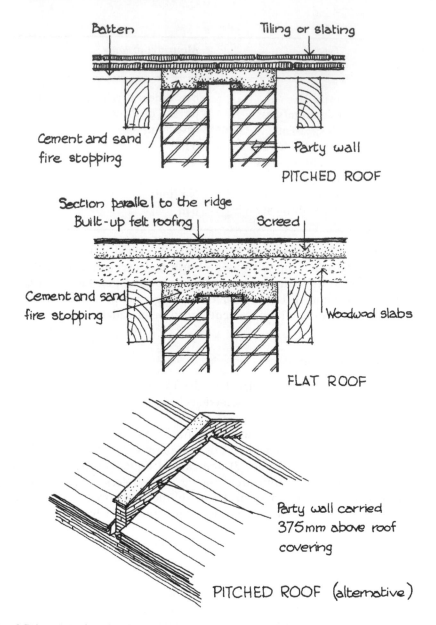

Fig. 6.5 Junction of roof and party wall

6.5.3 Ceilings

Section 6 of Approved Document B2 states that the ceiling of small rooms of not more than $4\,m^2$ in a residential building can be of a material with a surface spread of flame classification of Class 3. Other rooms and circulation spaces within dwellings should have ceilings which possess Class 1 grading and circulation space ceilings in the common areas of flats and maisonettes should be to Class O.

Gypsum plaster, plasterboard (painted or not or with a PVC facing not more than 0.5 mm thick), woodwool cement slabs and mineral fibre tiles or sheets with cement or resin binding all meet both Class O and Class 1 rating.

Timber boarding or plywood with a density more than $400\,\text{kg/m}^2$, painted or unpainted; wood particle board or hardboard either treated or painted and standard glass reinforced polyester products all are rated as Class 3. The timber products can be brought up to a Class 1 rating by an appropriate proprietary fire resisting treatment.

Other materials which may achieve the ratings shown but need to be substantiated by test evidence are:

Class O: aluminium faced fibre insulating board, flame retardent decorative laminates on a calcium silicate board, thick polycarbonate sheet and UPVC.

Class 1: phenolic or melamine laminates on a calcium silicate board and flame retardant decorative laminates on a combustible board.

6.6   Rooflights

Rooflights represent both a potential heat loss and a fire hazard. The heat loss is limited by including the area of any rooflights in with the windows in assessing the maximum permissible area of glazing which for single glazing is 15 per cent of the floor area and for double glazing is 30 per cent of the floor area.

The fire hazard aspect can be satisfied if the rooflight meets the requirements for a ceiling set out in Section 6.5.3 above. However, plastic rooflights with at least Class 3 rating may be used provided that they are at least 3 m apart and are 6 m from the boundary.

Chapter 7

INTERNAL WALLS

7.1   The Building Regulations applicable

A1   Loading
B2   Internal fire spread (linings)
B3   Internal fire spread (structure)
E1   Airborne sound (walls)
L1   Conservation of fuel and power
N1   Glazing

A1 requires that the building, in this case any loadbearing internal walls, must be constructed so that the loads are sustained and transmitted to the ground safely; B2 states that the internal linings, in this case the wall surfaces, must resist the spread of flame over their surface and if ignited have a rate of heat release which is reasonable in the circumstances; B3 has two requirements in relation to an internal wall: firstly, that in the event of a fire its stability will be maintained for a reasonable period and, secondly, that if it separates two or more buildings, i.e. semi-detached or terrace houses, flats or maisonettes, it must be designed and constructed to resist the spread of fire between the buildings; E1 is concerned with any wall that separates a dwelling from another dwelling or that separates a habitable room or kitchen within a dwelling from another part of the same building not used exclusively as part of the dwelling and states that it must resist the passage of airborne sound; L1 calls for the conservation of fuel and power, in this connection it applies to any wall between a dwelling and an unheated space such as a garage; N1 requires that any glazing with which people are likely to come into contact while in passage about the building will either, if broken, do so in a way which is not likely to cause injury or resist impact without breaking or be protected from impact.

7.1.1   Types of internal wall

The Building Regulations recognize a number of different types of internal wall depending on their particular function and apply appropriate standards to each. The types of wall are:

Separating walls Walls that divide one house, flat or maisonette from another or from another part of the building not used as part of the dwelling. These are also known as party walls but this is more of a legal reference to the ownership of the wall than to the wall function.

Compartment walls Walls that divide a building into fire resisting compartments to restrict the spread of a fire. In domestic work, any separating wall must also be constructed as a compartment wall.

Loadbearing wall Any wall that supports the floors or roof of the building.

Buttressing walls Walls that are built at right angles to another internal or external wall and provide it with structural restraint.

Partition walls Walls that subdivide the space enclosed by the external walls of the building but do not support any load other than the self-weight of the partition.

7.2 Structure

The Building Regulations require that compartment walls and partition walls have a substantial thickness, loadbearing walls can be thinner but still of sufficient substance to support the loads imposed, but partition walls, not being structural, are not the subject of any requirements under Regulation A1.

The minimum thickness for compartment walls is the same as that for external walls. This was given in Chapter 3 and is that the wall must be at least as thick as one-sixteenth of the storey height, subject to the further requirements shown in Table 3.2 and reproduced here as Table 7.1.

The rule for the minimum thickness for loadbearing walls given in Approved Document A1/2 is: half the thickness specified in Table 7.1 minus 5 mm, except for a loadbearing wall in the lowest storey of a three storey building which should have either the thickness given above or 140 mm, whichever is the greater. As an internal wall in a house or flat is hardly likely to exceed either 3.5 m high or 12 m long between supports, the effect of this rule is that the minimum thickness for a loadbearing wall generally is 90 mm and in the ground storey of a three storey building 140 mm or, in practical terms, walls of 100 mm blocks and 150 mm blocks respectively.

Whether it is a compartment wall or a loadbearing wall the compressive strength of the bricks or blocks should be:

- From the foundations to the first floor of of a three storey building   bricks $7\,\text{N/mm}^2$
                                       blocks $7\,\text{N/mm}^2$

- Any other internal wall   bricks $5\,\text{N/mm}^2$
                                             blocks $2.8\,\text{N/mm}^2$

Table 7.1 Minimum thickness of walls built of coursed brickwork or blockwork

Height of wall	Length of wall	Storey	Minimum thickness of a solid wall	Minimum thickness of the leaves of a cavity wall
Not exceeding 3.5 m	Not exceeding 12 m	All storeys	190 mm	90 mm each leaf
Exceeding 3.5 m but not exceeding 9 m	Not exceeding 9 m	All storeys	190 mm	90 mm each leaf
	Exceeding 9 m but not exceeding 12 m	Lowest storey*	290 mm	280 mm total but neither leaf to be less than 90 mm
		All upper storeys	190 mm	Each leaf 90 mm
Exceeding 9 m but not exceeding 12 m	Not exceeding 9 m	Lowest storey*	290 mm	280 mm total but neither leaf to be less than 90 mm
		All upper storeys	190 mm	90 mm each leaf
	Exceeding 9 m but not exceeding 12 m	Lowest two storeys*	290 mm	280 mm total but neither leaf to be less than 90 mm
		All upper storeys	190 mm	90 mm each leaf

(*Source:* Based on Table 5 of Approved Document A1 and paragraph 1C8)

* The height of the lowest storey is measured from the base of the wall.

7.2.1 Loading

The greatest span for any floor supported by a wall complying with the Approved Document guidance is 6 m measured from centre to centre of the bearing and the loading should be uniformly distributed along the wall. This can be assumed for concrete floors and for joisted timber floors with the joists at centres not exceeding 600 mm.

Lintels should bear 100 mm each end with spans under 1200 mm and 150 mm where the span is greater.

The wall should not be subject to any lateral loading and the combined dead and imposed load should not exceed 70 kN/m at the base of the wall.

7.3 Fire resisting walls

As already mentioned, the wall dividing a pair of semi-detached houses or a row of terrace houses and the walls dividing a building into flats or maisonettes must be constructed as compartment walls and possess the capacity to resist fire for the following periods:

- Houses   60 minutes
- Flats and maisonettes
  - up to 20 m high   60 minutes
  - 20 to 30 m high   90 minutes
  - over 30 m high   120 minutes
- Flat conversions (with adequate means of escape):
  - up to three storeys   30 minutes
  - four storeys and over   as for flats above

In addition the wall between an integral or attached garage and a house must have a minimum fire resistance of 30 minutes.

These standards can be achieved by a half brick wall without any form of finish in either a loadbearing or non-loadbearing wall: 100 mm lightweight and aerated concrete blocks without any finish will provide a fire resistance of up to 120 minutes in a loadbearing wall; in a non-loadbearing wall of lightweight blocks, a thickness of 50 mm is considered to be sufficient for 30 minutes protection and 75 mm for up to 120 minutes and, for aerated blocks, 50 mm is sufficient for up to 60 minutes and 63 mm for up to 120 minutes. In all cases the application of plaster to a wall makes insufficient difference to its fire resisting capacity to have any effect upon the practical thickness that can be built (but see Section 7.3.4 on wall linings and spread of flame).

The fact that the fire resisting requirement applied to a separating wall can be met by a half brick thickness or a 100 mm block does not mean that this is all that is required for this particular wall. As well as being fire resisting, a separating wall must also offer insulation against airborne sound and, in some cases, against heat loss. Neither of these standards can be met by a thin wall and Sections 7.4 and 7.5 should be consulted to find the actual thickness to be built.

7.3.1 Openings in fire resisting walls

Openings in compartment walls separating different occupancies should be limited to those for a door which is needed to provide a means of escape in case of fire and which has the same fire resistance as that required for the wall (see Chapter 8) or a pipe which must be properly fire stopped.

A door is also permitted in the wall between an attached or integral garage and a house. The door and its frame must be capable of resisting fire for 30 minutes and there must be a step up from the garage to the house of at least 100 mm.

7.3.2 Fire stopping

Pipes which pass through a compartment wall can be sealed with any proprietary system which has been shown to maintain the fire resistance of the wall; fire stopped, if the diameter is limited as shown below, or sleeved.

With a proprietary sealing system the diameter of the pipe is not restricted; if it is sleeved the maximum nominal internal diameter permitted is 160 mm and if it is to be fire stopped the maximum internal diameter depends on the material of the pipe as follows:

- Any non-combustible material   160 mm
- A stack pipe in lead, aluminium alloy
  PVC or fibre cement encased in a fire
  resisting duct   160 mm
- A branch pipe in lead, aluminium alloy
  PVC or fibre cement encased in a fire
  resisting duct   110 mm
- A stack pipe or branch pipe as specified
  above but not in a duct   40 mm
- Any pipe other than those above   40 mm

The fire stopping may be any proprietary material or cement mortar; gypsum based plaster; cement or gypsum based vermiculite/perlite mixes; glass fibre, crushed rock, blast furnace slag or ceramic based products (with or without resin binders), and intumescent mastics.

If a sleeve pipe is to be used it must be of non-combustible materials, such as steel or iron; it must project not less than 1 m on each side of the wall; it must be in contact with the service pipe and the opening through the wall must be as small as possible with fire stopping inserted between the structure and the sleeve pipe.

Where the service pipe is to be encased in a fire resisting duct, all the internal surfaces, including the wall face(s), are to have a spread of flame classification of Class O. The casing itself should:

1 have a fire resistance of not less than 30 minutes, including any access panels;
2 not be of sheet metal;
3 not have any access panels opening into a bedroom or circulation space;
4 be imperforate except for any openings for a pipe or access panel;
5 have the openings for pipes as small as possible and with fire stopping run around between the pipe and the structure of the duct.

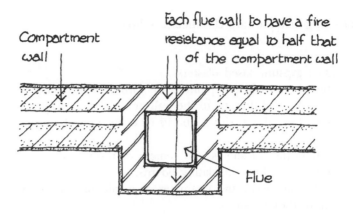

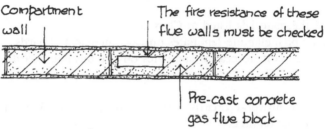

Fig. 7.1 Flues in compartment walls

7.3.3 Flues and ducts through fire resisting and compartment walls

Where a flue, a chimney containing a number of flues or a duct from an appliance passes through, or is built into a fire resisting wall each wall of the flue, chimney or duct should have a fire resistance equal to at least half that required for the main wall so that the standard of fire resistance is maintained (see Figure 7.1).

7.3.4 Wall linings

While the surface finish applied to a wall does not have any significant effect on its inherent fire resistance it does have an influence on the rate at which a fire can grow and spread through the building even though it is not usually the first material to be ignited. This is particularly important in circulation areas where a rapid growth and spread of the fire is most likely to prevent the occupants from escaping.

For this reason the common areas and corridors leading to flats and maisonettes should have a wall finish which is Class O whereas the finish to walls of circulation spaces and rooms within a dwelling should be Class 1 unless the room is under 4 m^2 when it can be Class 3. Parts of the wall finish in rooms may be of a lower classification than Class 1 provided that the total area does not exceed one-half of the floor area, subject to a maximum of 20 m^2.

Wall lining materials are classified as follows:

Class 0
- Cement or gypsum based plasters.
- Plasterboards, unpainted, painted or with a PVC facing not more than 0.5 mm thick fixed directly to the wall or with an air gap.
- Plasterboards as above with a fibrous or cellular insulating material behind.
- Ceramic tiling.
- Mineral fibre tiles or sheets with cement or resin binding.
- Aluminium faced fibre insulating boards.
- Flame retardant decorative laminates on a calcium silicate board.
- Thick polycarbonate sheet.
- Phenolic sheet.
- Unplasticized polyvinyl chloride materials.

Class 1
- Any of the materials listed as Class O
- Phenolic or melamine laminates on calcium silicate boards.
- Flame retardant decorative laminates on combustible boards.

Class 3
- Timber or plywood boards with a density of not less than $400 \, \text{kg/m}^3$, unpainted or painted.
- Wood particle boards or hardboard either treated or painted.

7.3.5 Sealing dry-lining

The infiltration of cold outside air through leakage paths significantly affects the space heating demand. To combat this, Approved Document L requires that any gaps between the walls and any dry-lining must be sealed. These can occur in the angles between the walls, at the junction between the walls and the floor or ceiling and around the edges of door and window openings. Sealing can be satisfactorily effected by continuous bands of fixing plaster.

7.3.6 Cavity barriers

Wherever a suspended ceiling is fixed below a floor a concealed cavity is formed which could give a path for the spread of fire and smoke through a building. While suspended ceilings are not very common in houses, they can be found in flats, particularly over corridors where they afford a convenient route for services.

To prevent a fire from bypassing the fire containment constructions of the building by running through such suspended ceiling voids, it is necessary to provide cavity barriers above and on the lines of any fire resisting walls. Compartment walls are of a higher standard of fire resistance than cavity barriers and must be carried right up to the underside of the floor.

Cavity barriers should:

1 provide a fire resisting integrity of 30 minutes and maintain its insulation for 15 minutes (not being structural there is no loadbearing requirement);
2 be tightly fitted to rigid construction and mechanically fixed in position;
3 not be rendered ineffective by movement of the building due to subsidence, shrinkage or temperature charge;
4 not be rendered ineffective due to movement of the building envelope due to wind pressure;
5 not collapse due to a fire in any services penetrating the barrier;
6 not fail because their fixings could not withstand fire.

7.4 Sound insulation

The wall which separates semi-detached houses, one terrace house from its neighbour or adjoining flats or maisonettes must be capable of keeping the noise of domestic activities in the adjoining property down to a level that will not threaten the health of the occupants of the dwelling and will allow them to sleep, rest and engage in their domestic activities in satisfactory conditions. The sound to be dealt with is airborne sound.

Approved Document E1 shows four types of wall considered satisfactory: a solid masonry wall, a cavity masonry wall, a masonry wall between isolated panels and a timber framed wall with absorbency material.

The solid masonry wall shown in Fig. 7.2 depends on its mass to absorb the sound – being heavy it is not so easily set into vibration. The constructions given will achieve the recommended masses which, in the case of the brick wall should be $375 \, \text{kg/m}^2$ including the finish and the concrete blocks and *in situ* concrete should be $415 \, \text{kg/m}^2$. The junction with an external wall shows both a masonry and a timber framed inner skin with an outer leaf of any construction. Where the inner leaf is of masonry (or the external wall is solid) the walls should be bonded or tied together with ties at not more than 300 mm vertically and the external wall should have a mass of not less than $120 \, \text{kg/m}^2$ unless its length is limited by openings which:

1 are not less than 1 m high;
2 are on both sides of the separating wall;
3 are not more than 700 mm from the face of the separating wall on both sides.

Where the inner leaf is of timber framing it should be tied to the separating wall with ties at not more than 300 mm apart vertically and the joints should be sealed with tape or caulked.

The cavity wall and the wall with isolated panels depend partly on their mass and partly on the structural isolation between the leaves.

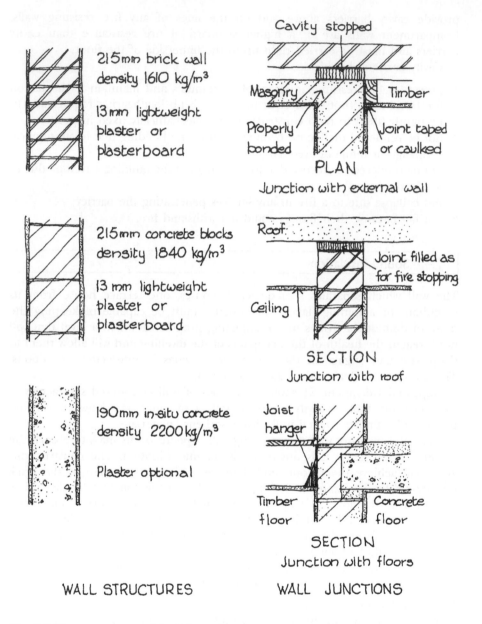

Fig. 7.2 Mass masonry sound insulating walls

Figure 7.3 shows typical constructions of cavity walls for sound insulating purposes. The total mass of the brick and the concrete block, including the finish, should be not less than $415 \, \text{kg/m}^2$ and for the lightweight block wall should be not less than $300 \, \text{kg/m}^2$. Where there is a ceiling of 12.5 mm plasterboard or equivalent, the mass of the wall above the line of the ceiling may be reduced to $150 \, \text{kg/m}^2$. The cavity should still be maintained. If

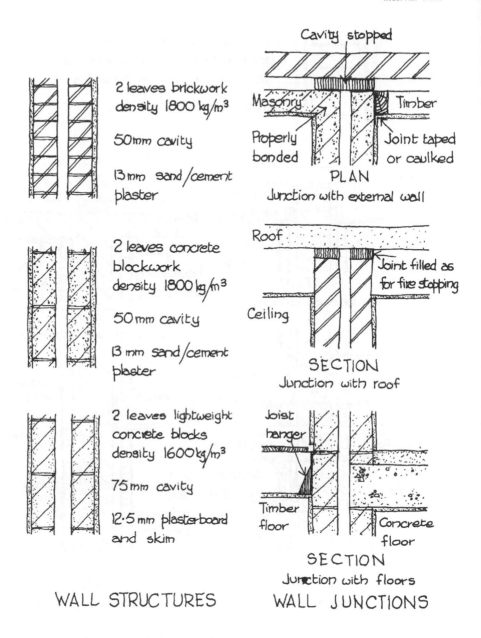

WALL STRUCTURES

2 leaves brickwork density 1800 kg/m³

50mm cavity

13mm sand/cement plaster

2 leaves concrete blockwork density 1800 kg/m³

50mm cavity

13 mm sand/cement plaster

2 leaves lightweight concrete blocks density 1600 kg/m³

75mm cavity

12·5 mm plasterboard and skim

WALL JUNCTIONS

Cavity stopped

Masonry   Timber

Properly bonded   Joint taped or caulked

PLAN
Junction with external wall

Roof

Joint filled as for fire stopping

Ceiling

SECTION
Junction with roof

Joist hanger

Timber floor   Concrete floor

SECTION
Junction with floors

Fig. 7.3 Cavity masonry sound insulating walls

lightweight aggregate blocks with a density less than 1200 kg/m³ are used, one face should be sealed with cement paint or plaster skim. The leaves should be tied together with butterfly wire wall ties as required for structural purposes and the cavity kept clear. This means that if the external wall cavity is to be filled with a loose type of thermal insulation material some means must be used to prevent it from entering the separating wall cavity, such as by the use

113

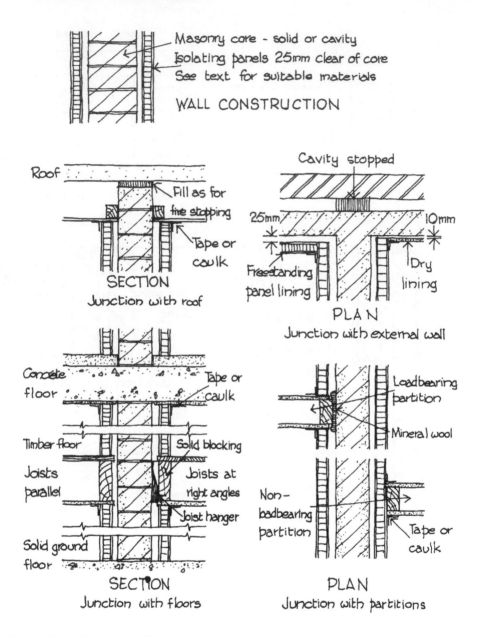

Masonry core - solid or cavity
Isolating panels 25mm clear of core
See text for suitable materials

WALL CONSTRUCTION

Roof

Fill as for
fire stopping

Tape or
caulk

SECTION
Junction with roof

Cavity stopped

25mm   10mm

Freestanding
panel lining

Dry
lining

PLAN
Junction with external wall

Concrete
floor

Tape or
caulk

Timber floor

Solid blocking

Joists
parallel

Joists at
right angles

Joist hanger

Solid ground
floor

SECTION
Junction with floors

Loadbearing
partition

Mineral wool

Non-
loadbearing
partition

Tape or
caulk

PLAN
Junction with partitions

Fig. 7.4 Sound insulating wall of masonry core and isolated panels

of cavity batts in the external wall at the junction. The connection with the external wall should be constructed in the same way as a solid sound insulating wall.

The wall with a masonry core and isolated lining panels, shown in Fig. 7.4, can have a number of combinations of core and panel. Approved Document E defines four cores and two panels. The four cores are brick with a mass of

$300 \, \text{kg/m}^2$ such as $215 \, \text{mm}$ of brickwork with a density of $1290 \, \text{kg/m}^3$; concrete block, also with a mass of $300 \, \text{kg/m}^2$, which can be achieved by a $140 \, \text{mm}$ wall of blocks of a density of $2200 \, \text{kg/m}^3$; lightweight concrete block with a mass of $160 \, \text{kg/m}^2$ which could be a $200 \, \text{mm}$ thickness of blocks of a density of $730 \, \text{kg/m}^3$; the fourth is a cavity core and can be of any mass but the leaves should be at least $100 \, \text{mm}$ thick with a cavity of $50 \, \text{mm}$ and tied only with butterfly wire wall ties.

The panels can be either two sheets of plasterboard joined together by a cellular core which should have a mass, including any applied plaster finish, of not less than $18 \, \text{kg/m}^2$ and the joints between the panels must be taped or two sheets of plasterboard with staggered joints. If a supporting framework is used for the double layer of plasterboard each can be $12.5 \, \text{mm}$ thick but if it has no framework the finished thickness must be at least $30 \, \text{mm}$.

The junction with an external wall depends on the material of the core. A lightweight block core requires the external wall to be lined with free standing lining panels similar to the separating wall, where the core is of any of the other three types, the inner skin of the external wall may be plastered or dry lined as shown but must have a total mass of not less than $120 \, \text{kg/m}^2$ and be butt jointed to the separating wall with ties at no more than $300 \, \text{mm}$ vertically.

Where this type of separating wall supports a suspended timber floor the joists should be carried on joist hangers and solid blocking fixed between them on the line of the isolated panel. Joists running parallel to the wall should be positioned to maintain the air space between the isolated panels and the core. A concrete suspended floor can only be run through the core as shown if its mass is at least $365 \, \text{kg/m}^2$ (if it is less the core wall must run through) and if the core is of cavity form the floor must not bridge the cavity.

Partitions which abut this form of separating wall should not be of masonry construction and should be built as shown with loadbearing partitions secured to the core wall but separated from it by a continuous pad of mineral wool and non-loadbearing partitions fixed to the isolating panels only.

Timber framed walls rely on isolation and insulation for their sound insulating properties and a typical section through such a wall is shown in Fig. 7.5. The construction is of two independent stud frames faced with two layers of $15 \, \text{mm}$ plasterboard and with a $25 \, \text{mm}$ quilt of mineral wool suspended between the frames. It is essential with this system that, at no point, is there any firm contact between the two halves of the wall because that would allow a direct transference of sound energy vibrations from one to the other and completely negate the insulating properties.

There are no restrictions on the junction between this type of sound insulating wall and a timber framed external wall but if the wall is of cavity masonry construction the cavity must be sealed between the ends of the separating wall and the outer leaf to prevent flanking transmission sound paths being formed.

It should be noted that this construction will also provide a fire resistance of one hour.

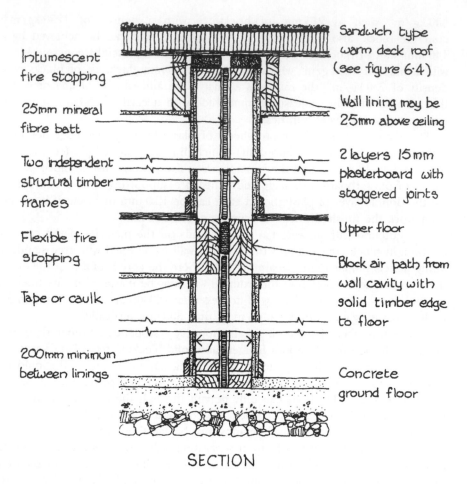

Intumescent
fire stopping

Sandwich type
warm deck roof
(see figure 6·4)

25mm mineral
fibre batt

Wall lining may be
25mm above ceiling

Two independent
structural timber
frames

2 layers 15mm
plasterboard with
staggered joints

Upper floor

Flexible fire
stopping

Block air path from
wall cavity with
solid timber edge
to floor

Tape or caulk

200mm minimum
between linings

Concrete
ground floor

SECTION

Fig. 7.5 Typical sound insulating fire resisting timber framed partition

7.5 Thermal insulation

The majority of internal walls are not affected by considerations of conservation of fuel and power but there are a few particular situations where this must be observed. In principle, any wall which divides a heated space or room from another which is neither heated nor insulated to the standards laid down must restrict the amount of heat passed from one side of the wall to the other. This situation can occur with the wall between a house and an attached or integral garage, or in flats with the wall between the flats and the stairways or corridors where the structure enclosing the corridor or stair does not have the U value applicable to the rest of the building (see Chapter 3, Section 3.4 for an explanation of 'U value').

Such walls are known as 'semi-exposed' and should have a U value of

Table 7.2 Thicknesses of insulation (mm) for a design U values of 0.45 W/m²K and 0.6 W/m²K

Insulation material		Phenolic foam		Polyurethane board		Expanded polystyrene slab, Glass fibre slab, or Mineral fibre slab		Glass fibre quilt, or Urea formaldehyde foam	
Thermal conductivity	(W/mK)	0.020		0.025		0.035		0.040	
Design U value of wall	W/m² K	0.45	0.6	0.45	0.6	0.45	0.6	0.45	0.6
Base thickness of insulation	mm	41	30	51	37	71	52	82	59

Allowable reductions for wall components:

Brick outer leaf	2	3	4	5
100 mm Concrete outer leaf: kg/m³				
Aerated 600	8	10	15	17
800	7	8	12	14
Lightweight 1000	6	7	10	11
Dense 1600	3	3	5	5
100 mm Concrete inner leaf: kg/m³				
Aereated 600	9	11	15	17
800	7	9	13	15
Lightweight 1000	6	8	11	12
Dense 1600	3	4	5	6
50 × 100 nom. Timber frame and glass fibre slab	42	53	74	84
50 × 100 nom. Timber frame and glass fibre quilt	38	48	67	77
Cavity (25 mm minimum)	4	5	6	7
13 mm plaster	1	1	1	1
13 mm lightweight plaster	2	2	3	3
10 mm plasterboard	1	2	2	3
13 mm plasterboard	2	2	3	3
Airspace behind plasterboard dry lining	2	3	4	4
9 mm sheathing ply	1	2	2	3
20 mm cement render	1	1	1	2
13 mm tile hanging	0	0	1	1

(Source: Tables A5, A6, A7 and A8 of Appendix A to Approved Document L: 1995 Edition)

0.6 W/m²K. The means of achieving this thermal insulation standard is to assess the U value of the structure and calculate the thickness of insulation to be added.

Approved document L provides four tables to make this process easier. These were used as the basis for Table 3.3 which, for convenience, is reproduced here as Table 7.2. To find the insulation thickness required the material to be used must first be selected. Below this can be found the design U value of $0.60 \text{ W/m}^2\text{K}$ and below that the base thickness of the insulation. The allowable reductions for the components of the wall are deducted from this base thickness to find the minimum thickness needed. The reduction figures for a brick leaf or a concrete inner leaf can be used for a brick or a block partition. Once the minimum figure has been established, it may be necessary to round it up to the next available greater thickness. If this is not commercially available the nearest thickness above should be used. The following list gives the average thermal conductivity of insulation materials.

Material	Thermal conductivity (W/mK)
Polyurethane board	0.025
Expanded polystyrene slab (EPS)	0.035
Glass fibre slab	0.035
Mineral fibre slab	0.035
Glass fibre quilt	0.040
Phenolic foam	0.020
Urea formaldehyde (UF) foam	0.040

The thermal insulation value of lightweight concrete blocks relates to the density of the block. Typical values are:

Block density (kg/m³)	Thermal conductivity (W/mK)
500	0.16
600	0.19
700	0.21
800	0.23

Figure 7.6 shows a number of possible constructions which satisfy this requirement.

7.6 Glazing in internal walls

Any internal walls which contain glazed panels down to floor level are considered to be a hazard unless certain criteria are met. The critical location for such glazing is between finished floor level and 800 mm above the floor. This height is increased to 1500 mm for a distance of 300 mm on each side of a door (and in the door itself, if it is glazed).

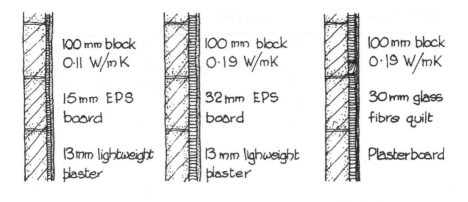

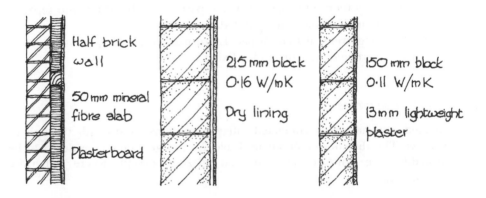

All constructions provide at least 30 minutes fire resistance
and a U value of at least 0.60 W/m²K

Fig. 7.6 Constructions for walls between a house and a garage

The risk can be reduced in these locations if the glazing: breaks safely – if it breaks; is robust or in small panes; or is permanently protected.

7.6.1 Safe breakage

In practice, safe breakage is concerned with the performance of laminated and toughened glass and is defined in BS 6206: *Specification for impact performance requirements for flat safety glass and safety plastics used in building*: Clause 5.3. The requirement of the BS is that safety glass and plastics should either 'not break' or 'break safe'. The latter meaning that if it does break the impact results in

Table 7.3 Maximum areas of safety glass (m^2)

Thickness (mm)	Annealed glass	Toughened or laminated glass
3	0.2	Not recommended
4	0.3	Not recommended
5	0.45	1.0
6	0.7	3.0
10	1.5	6.0
12	3.0	7.0

either a small clear opening only, with a limit to the size of the detached particles; disintegration into small detached particles or breakage into separate pieces that are neither sharp nor pointed.

It is also a requirement that all panels are marked as safety glass in such a way that it is permanently visible after fixing.

7.6.2 Robustness

Annealed, toughened or laminated safety glass gains its strength through its thickness. The thicknesses shown in Table 7.3 are generally considered to be reasonably safe for ordinary internal glazing except where otherwise recommended.

7.6.3 Glazing in small panes

In this context, a 'small pane' may be a single isolated pane of glass or a number of panes contained within glazing bars.

Approved Document N1 recommends that the width of any small pane should be not more than 250 mm and its area not more than 0.5 m^2. These measurements are to be taken between glazing beads or similar fixings. As can be seen from Table 7.3 the minimum thickness for annealed glass in a small pane of this size is 6 mm except in traditional leaded or copper-light glazing in which 4 mm glass would be acceptable, provided that there is no requirement for fire resistance.

7.6.4 Permanent screen protection

As an alternative to the provision of safety glass or the subdivision of the glazed area into small panes, the glazing can be provided with a suitable form of permanent protection on both sides. The protective screen should be designed and constructed so that:

1 it is 800 mm high above the finished floor level;
2 a sphere of 75 mm diameter cannot come into contact with the glass;
3 it is sufficiently robust to afford adequate protection;
4 if the partition containing the glazed area is intended to provide a guard against falling, such as the enclosure at the head of a staircase, the protective screen should not have horizontal bars or any other design feature which would make it climbable (see Fig. 7.7).

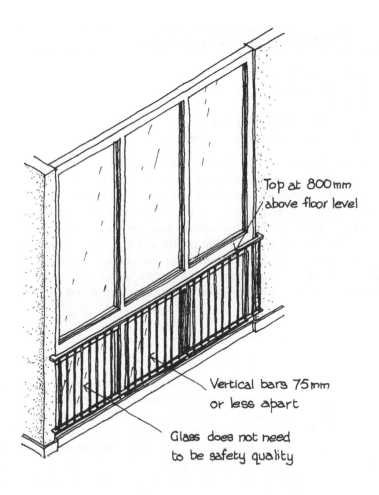

Fig. 7.7 Permanent protection for wall glazing

Chapter 8

WINDOWS, DOORS AND VENTILATION OPENINGS

8.1   The Building Regulations applicable

B4   External fire spread
F1   Means of ventilation
J1   Air supply
L1   Conservation of fuel and power
N1   Glazing – materials and protection

B4 requires that the external walls (including the windows and doors) resist the spread of fire over the walls and from one building to another; F1 requires that there be adequate means of ventilation provided for people in the building; J1 is concerned that there is an adequate supply of air to heat producing appliances for combustion and for the efficient working of the chimney; L1 states that reasonable provision be made for the conservation of fuel and power in buildings; and N1 is concerned that glazing with which people are likely to come into contact while in passage about the building, should either break safely, resist breaking or be protected from impact. **D8.1**

8.2   Windows

As one of the principle sources of heat loss, the windows in a building must receive special attention. Approved Document L states that the standard U value for windows in a dwelling should be 3.0 W/m^{2}K if the Energy Rating is 60 or less and 3.3 W/m^{2}K if it is over 60 (see Chapter 1, Sections 1.9 and 1.10 for explanations of Energy Rating). The latter figure of 3.3 W/m^{2}K also applies to extensions to existing dwellings but, in this case, the basic allowance can be applied either to the floor area of the extension or to the floor area of the dwelling and the extension taken together.

These figures are a basic allowance and can be used to prove compliance with the requirements of the Regulations by the Elemental Method (see Section 1.10.1). They represent the average of the U value of the doors and any rooflights as well as the windows and the requirements will only be met if the area of the windows, doors and rooflights, taken together, does not exceed 22.5 per cent of the total floor area of the house or flat.

If the design calls for a larger proportion of window than is permitted in the basic allowance, the Target U value Method would provide a means of demonstrating that the proposals can be made capable of meeting the requirements. It would be necessary, of course, to provide either windows of a better thermal performance than the figures given above or some other part of the structure must be given a superior thermal resistance. Conversely, if the thermal performance of the window is inferior to the standard, it can be used but its area must be restricted. Table 8.1 shows the permitted variations in the area of doors and windows for differing U values.

Table 8.1 Permitted variation in the total area of windows and doors

Average U value (W/m^2K)	Maximum permitted total area of windows and doors as a percentage of the floor area	
	SAP 60 or less	SAP over 60
2.0	37.0%	41.5%
2.1	35.0%	39.0%
2.2	33.0%	36.5%
2.3	31.0%	34.5%
2.4	29.5%	33.0%
2.5	28.0%	31.5%
2.6	26.5%	30.0%
2.7	25.5%	28.5%
2.8	24.5%	27.5%
2.9	23.5%	26.0%
3.0	**22.5%**	25.0%
3.1	21.5%	24.0%
3.2	21.0%	23.5%
3.3	20.0%	**22.5%**
3.4	19.5%	21.5%
3.5	19.0%	21.0%
3.6	18.0%	20.5%
3.7	17.5%	19.5%
3.8	17.0%	19.0%
3.9	16.5%	18.5%
4.0	16.0%	18.0%
4.1	15.5%	17.5%
4.2	15.5%	17.0%

(*Source:* Table 3 of Approved Document L: 1995 Edition)

Notes: 1. The table assumes that all the walls, floors and roof possess the standard U values given in 1.10.1.
2. This table also assumes that there are no rooflights in the dwelling.

8.2.1 Window constructions

The frame of the window affects the thermal performance as well as the glazing with which it is fitted. The traditional timber frame offers one of the best insulation standards and simple metal frames with their high rate of thermal transmission give the worst. To alleviate the problem with metal frames, designers have introduced a thermal break into the frame that effectively divides the frame into inner and outer components, separated by a thermally insulating material. The newer PVC-U window frames are manufactured to have almost the same thermal resistance as timber, as can be seen from the figures in Table 8.2.

Most window frames now need to be fitted with one of the forms of insulated glazing. The simplest is the sealed double glazing unit made up with either normal float glass or, if the size of the pane warrants it, toughened glass. The width of the space between the sheets of glass determines the degree of resistance of the unit. The optimum is in the region of 20 mm. Less than this and heat can radiate across the gap, more than this and convection currents can be set up to convey the heat from one sheet of glass to the other. In practice, a smaller gap is often used to avoid excessively wide frames. The smallest commonly found to be effective is 6 mm but 12 mm is also available and gives an improved performance.

Generally, the gap is filled with dry air but replacing this with argon gas increases the resistance to thermal transmission.

A more recent development has been the production of glasses with a low emissivity (low-E). These allow the sun's rays to pass into the building but restrict heat passing out.

Table 8.2 Indicative U values (W/m^2K) for windows and rooflights

Type of glazing	Type of frame							
	Wood		Metal		Thermal break		PVC-U	
Single glazed, normal glass	4.7		5.8		5.3		4.7	
Double glazed units with a width of gap of: (mm)	6	12	6	12	6	12	6	12
Double glazed, normal glass	3.3	3.0	4.2	3.8	3.6	3.3	3.3	3.0
Double glazed, low-E glass	2.9	2.4	3.7	3.2	3.1	2.6	2.9	2.4
Double glazed, normal glass, Argon filled	3.1	2.9	4.0	3.7	3.4	3.2	3.1	2.9
Double glazed, low-E glass, Argon fllled	2.6	2.2	3.4	2.9	2.8	2.4	2.6	2.2
Triple glazed, normal glass	2.6	2.4	3.4	3.2	2.9	2.6	2.6	2.4
Rooflights at less than 70 deg. from horizontal Double glazed, normal glass	3.6	3.4	4.6	4.4	4.0	3.8	3.6	3.4

(*Source:* Table 2 of Approved Document L: 1995 Edition)

The U value of a window and its glazing can be found in some Manufacturer's literature but, if this is not available, the values given in Table 8.2 can be used. From this it can be seen that the required standards for windows cannot be achieved with single glazing.

There is one relaxation of the rule. A single glazed window may be used where it is protected by an unheated, enclosed, draught-proof porch or conservatory, in which case it can be assumed to have a U value of 3.3 W/m^{2}K.

8.2.2 Draught sealing

Approved Document L points out that space heating demand is significantly affected by the unintentional infiltration of cold outside air. Some ventilation of the building is essential for the health of the occupants and is dealt with in Section 8.4 but it must be controlled so that it is not excessive and is acceptably located.

To achieve this, the Approved Document requires that all openable elements of windows are fitted with draught-stripping and that the junction between the frame and the window reveal is suitably sealed all round.

8.2.3 Solar gain

Windows can be a source of heat gain as well as a cause of heat loss. The sun's rays can pass through glazing and heat up the objects within the room onto which they fall, thereby warming the room. This is known as 'solar gain'.

Advantage can be taken of this effect if the Target U value Method is used to assess the thermal performance of the dwelling (see 1.10.2). It is assumed in the equations used in this method that the glazing is equally distributed on the North and South elevations. If this is not the case and there is a greater area of window facing South than facing North, 40 per cent of the difference between the two can be subtracted from the total actual window area when calculating the heat losses. The calculation shown in 1.10.2 demonstrates this in use.

8.3 External doors

As well as windows, **D8.2** the external doors of a building need to be considered from the point of view of thermal loss and added to the heat losses through the windows when assessing the overall thermal performance of the building.

A solid timber door affords the best resistance to heat loss and is the only form possible if the Energy rating of the dwelling is 60 or less. Half or fully glazed doors using double glazed units are acceptable where the Energy rating is over 60. Half glazed doors using single glazing could be fitted if there is a compensating improvement in the thermal resistance of some other element in the fabric but a door with a full length panel of a single sheet of glass is not likely to be acceptable. Table 8.3 shows indicative U values for doors. If certified manufacturer's data is available it should be used in preference to the figures given in the table. **D8.3**

8.3.1 Draught sealing

As is explained in Section 8.2.2, uncontrolled infiltration of cold outside air should be avoided. Doors and their frames should receive the same treatment as windows, i.e. the frame should be fitted with draught-stripping and it should be sealed all round its perimeter where it abuts the wall reveal. **D8.4**

8.4 Ventilation

In general, ventilation has three purposes:

1 To extract moisture from areas such as kitchens and bathrooms where it is produced in significant quantities.
2 To achieve occasional rapid ventilation for the dilution of pollutants and of moisture likely to produce condensation in habitable rooms and sanitary accommodation.
3 To achieve secure and comfortable background ventilation in habitable rooms and kitchens.

In addition, Regulation J1 requires that there is an adequate supply of air to heat producing appliances for the purposes of combustion and efficiency in the working of any chimney.

The smallest dimension of any ventilation opening should be not less than

Table 8.3 Indicative U values (W/m^2K) for doors

Type of door	Type of frame							
	Wood		Metal		Thermal break		PVC-U	
Solid timber panel	3.0		–		–		–	
Single glazed, normal glass	4.7		5.8		5.3		4.7	
Half single glazed, half timber panel	3.7		–		–		–	
Double glazed units with a width of gap of: (mm)	6	12	6	12	6	12	6	12
Fully double glazed, normal glass	3.3	3.0	4.2	3.8	3.6	3.3	3.3	3.0
Half double glazed, normal glass	3.1	3.0	3.6	3.4	3.3	3.2	3.1	3.0

(*Source:* Table 2 of Approved Document L: 1995 Edition)

8 mm to minimize resistance to the flow of air. This does not apply to a screen, fascia, baffle, etc.

All ventilation openings, whether a window to provide rapid ventilation or a trickle vent or air brick that provides background ventilation should be s'.ed so as to avoid discomfort due to cold draughts. This, it is suggested in Approved Document F1, should be 1.75 m above the floor level. They should also be placed so as to prevent the ingress of rain.

Background ventilation provisions should be reasonably secure and they should be adjustable. **D8.5**

8.4.1 Definitions

To be clear in this context what is meant by the terms used, the following definitions have been adopted:

Ventilation opening: any means of ventilation, permanent or closable, which opens directly to external air. This includes the openable parts of a window, an external door, a louvre, an air brick, or a window trickle ventilator.

Passive stack ventilation (PSV): a ventilation system using ducts from the room ceilings to terminals on the roof that operates by a combination of the natural stack effect and wind blowing over the roof. The natural stack effect is the phenomenon that occurs in a chimney and is the natural movement of air up a flue or duct due to the difference in density between the internal air and the cooler outside air.

Habitable room: any room used for dwelling purposes but not a kitchen.

Bathroom: any room containing either a bath or a shower.

Sanitary accommodation: any room containing a W.C.

The particular ventilation requirements relate to the specific use to which the room is to be put as shown below and illustrated in Fig. 8.1.

8.4.2 Generally

The following clauses set out the ventilation requirements for each individual room. An alternative approach is to provide a level of background ventilation throughout the building that is equivalent to an average of 6000 mm^2 per room for the rooms listed below, with a minimum provision of 4000 mm^2 in each room.

Where a room serves a dual function, such as a kitchen-diner, the individual provisions do not need to be duplicated. However, the greater of the individual provisions for rapid, background and extract ventilation must be installed.

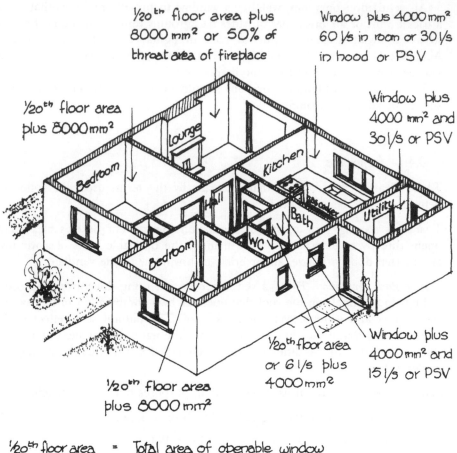

1/20th floor area plus 8000 mm² or 50% of throat area of fireplace

Window plus 4000 mm² 60 l/s in room or 30 l/s in hood or PSV

1/20th floor area plus 8000 mm²

Window plus 4000 mm² and 30 l/s or PSV

Lounge

Bedroom

Kitchen

Hall

Bath

Utility

WC

Bedroom

1/20th floor area or 6 l/s plus 4000 mm²

Window plus 4000 mm² and 15 l/s or PSV

1/20th floor area plus 8000 mm²

1/20th floor area	=	Total area of openable window
Window	=	Openable window - no min. size
8000 mm²	=	Area of trickle vent or air brick
30 l/s	=	Extract fan rate in litres per second
PSV	=	Passive stack ventilation

Fig. 8.1 Ventilation requirements

Mechanical ventilation is required in some rooms but should not be installed in any room where there is a solid fuel appliance. In this case, advice should be sought from the Heating Equipment Testing and Approval Scheme (HEATS), PO Box 37, Bishop's Cleeve, Gloucester, GL52 4TB.

8.4.3 Habitable rooms

A habitable room must have two forms of ventilation: rapid ventilation of at least one-twentieth of the floor area and background ventilation of not less than 8000 mm^2.

The rapid ventilation can, and usually does, comprise an opening window or windows of which some part must be at least 1.75 m above the floor but it would also include any hinged French doors, sliding patio doors or similar external openings provided that they satisfy the height requirement.

It is increasingly common to find that an external door to a habitable room which provides this rapid ventilation opens into a conservatory or similar space. This is permissible provided that the following conditions are met:

1 There is an opening between the room and the conservatory with an area of not less than one-twentieth of the combined floor areas of the room and conservatory. This area may be fixed open or may be closable. As it is very probable that the room and the conservatory will be connected by one, if not a pair of doors, this rule is not difficult to meet.
2 There are one or more rapid ventilation openings with a total area of not less than one-twentieth of the combined floor areas of the room and the conservatory of which some part is at least 1.75 m above the floor.
3 There are background ventilation openings in the conservatory of a total of 8000 mm^2 and similar openings of the same size between the conservatory and the room. These background vents must be located so that they do not cause undue draughts. If the opening between the two rooms is permanently open they can be treated as one space and the requirement of 8000 mm^2 background ventilation only applies to the conservatory or outer room.

8.4.4 Kitchens

Rapid ventilation in the form of an opening window is required in a kitchen but no minimum size is given for this window. In addition, background ventilation of 4000 mm^2 and extract ventilation are required. The latter can either be mechanical or by means of passive stack ventilation.

Mechanical extraction should be at the rate of 30 l/s where the fan is in a cooker hood or located near the hob and controlled by a humidistat. By 'near the hob' is meant adjacent to the ceiling and within 300 mm of the centre line of the hob. If the extract fan is sited elsewhere in the kitchen its capacity should be increased to 60 l/s. Mechanical extract ventilation should not be installed in any room where there is a solid fuel appliance. It probably would not be necessary in this case as it is likely that the chimney would provide adequate passive stack ventilation.

Passive stack ventilation is considered adequate if it is to an open-flued appliance that is the primary source of heating, cooking or hot water production, or the flue is equivalent to a 125 mm diameter duct and there are no control dampers that could block the free flow of air. Alternatively it should be provided in accordance with BRE Information paper 13/94 or be supported by suitable certification such as a BBA Certificate. Further advice on this can be obtained from the Heating Equipment Testing and Approval Scheme (HEATS), PO Box 37, Bishop's Cleeve, Gloucestershire, GL52 4TB.

Both the mechanical ventilation and the passive stack ventilation can be operated manually or automatically by sensor or controller.

If the kitchen is an internal room, i.e. does not have an outside wall, or if it cannot have any opening windows, any mechanical ventilation provided should have a 15 minute over-run and be controlled either automatically or via the light switch. Alternatively it may be ventilated by a PSV or an open-flued appliance and, in all cases it must be provided with an air inlet such as a 10 mm gap under the door.

8.4.5 Bathrooms

Where the Bathroom is located on an outside wall, an opening window (of no particular minimum size) and background ventilation of $4000 \, mm^2$ are required. In addition, extract ventilation either by mechanical means with a 15 l/s capacity or PSV is to be provided.

If the room is not on an external wall, any extract fan should have a 15 minute over-run and the room be provided with an air inlet such as a 10 mm gap below the door.

8.4.6 Sanitary accommodation

Both rapid ventilation and background ventilation are required in a WC. The rapid ventilation can be either by means of an extract fan with a capacity of 6 l/s and an over-run of 15 minutes, or an openable window equal to 1/20th of the floor area. This can readily be achieved in the average WC by, for instance, a small night vent in a window 450 mm wide.

The background ventilation should be equal to $4000 \, mm^2$.

8.4.7 Utility rooms

A utility room with access only from the outside requires no special provision for ventilation. One accessible from the inside of the building (but on an outside wall) must be provided with an opening window, of no particular minimum size, $4000 \, mm^2$ background ventilation and extract ventilation in the form of either a fan capable of 30 l/s or PSV. If the utility room is an internal room, its ventilation requirement is an extract fan that is capable of 30 l/s with an overrun of 15 minutes controlled either automatically or manually via the light switch, or, as an alternative to a fan, passive stack ventilation can be installed. In both cases, an air inlet is required which could take the form of a 10 mm gap below the door.

8.4.8 Air supply to heat-producing appliances

As well as ventilation for the comfort of the occupants, ventilation openings are also required for the benefit of any solid fuel, gas or oil burning appliance unless it has a balanced flue system which allows combustion air in as well as taking the products of combustion out.

Different fuels and the type and size of appliance each require a varying amount of air supply. These are summarized in Table 8.4.

8.5 Glazing

Section 7.6 of the previous chapter gave details of the precautions to be taken against glass in vulnerable positions in internal walls causing injury by accidental breakage. The same rules apply to glazing in windows and doors.

The critical location for glazing is within 800 mm of the floor level and would apply to the lower part of a feature window or a glazed door. In this position the glazing, if hit, should either not break or break safely. Alternatively, permanent protection can be provided to prevent the glazing being hit. Prevention of breakage can be achieved by using impact resistant materials or by restricting the size of the panes to a maximum width of 250 mm and a maximum area of 0.5 m^2.

Safe breakage is when impact with the glazing results in only a small clear hole and detached particles of limited size, disintegration into small detached particles or shattering into pieces that are neither pointed nor sharp.

The safe size for annealed and laminated safety glazing of various thicknesses is given in Table 7.3 and is reproduced here for convenience as Table 8.5.

Where fire resistance is not a controlling factor, 4 mm glass is acceptable in traditional leaded glazing or copper-lights.

Permanent screen protection is an option more suited to the lower part of windows than to glazed doors but, if provided, removes any need to have safe glazing. The screen should be robust and designed so that a sphere of diameter 750 mm cannot pass through at any point to make contact with the glass. If the glazing protects against a danger of falling, the screen should be difficult to climb (see Fig. 7.7).

8.6 Spread of fire

Chapter 3 deals with external walls and the need for them to prevent the spread of fire from one building to another. Windows and doors facing a

Table 8.4 Ventilation openings to provide an air supply to heat producing appliances (other than balanced flue units)

Fuel	Type of appliance	Openings with a total free area of:
Solid fuel	Open	50% throat opening
	Any other	550 mm^2 for each kW of rated output over 5 kW. With a draught stabilizer this must be increased by 300 mm^2 for each kW of output
Gas	Cooker	Openable window or other means of ventilation of any size. If the room volume is less than 10 m^3 a permanent vent of 5000 mm^2 is needed
	Open-flued	450 mm^2 for each kW of input over 7 kW
Oil	Any	550 mm^2 for each kW of ouput above 5 kW

(*Source:* Based on Approved Document J1/2/3)

Table 8.5 Maximum areas of safety glass (m^2)

Thickness (mm)	Annealed glass	Toughened or laminated glass
3	0.2	Not recommended
4	0.3	Not recommended
5	0.45	1.0
6	0.7	3.0
10	1.5	6.0
12	3.0	7.0

boundary present a weakness in this fire protection and these unprotected areas must be restricted.

Very small openings or small openings which are well apart do not present a hazard and can be ignored. The permitted limits for these small openings is shown in Fig. 3.7, reproduced here as Fig. 8.2.

Beyond these small openings, the total area of windows and doors in the side of a house or block of flats which is neither over three storeys high nor

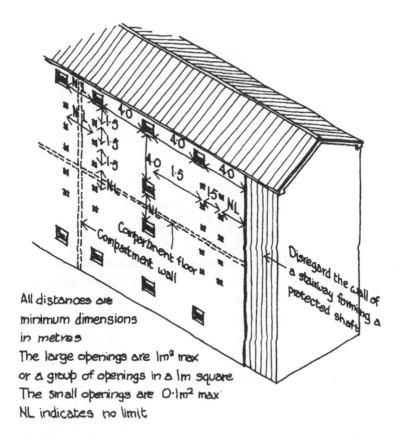

All distances are
minimum dimensions
in metres

The large openings are 1m² max
or a group of openings in a 1m square

The small openings are 0·1m² max

NL indicates no limit

Fig. 8.2 Unprotected areas which may be disregarded for separation distance purposes

more than 24 m long should fall within the limits shown in Fig. 8.3 in relation to its minimum distance from any boundary which is at an angle of less than 80° to the wall.

Windows and doors in houses and flats which are more than three storeys high or 24 m long should conform to the following limits:

Distance from the building to the boundary (m)	Maximum total area as a percentage of the wall area
1	4%
2.5	8%
5	20%
7.5	40%
10	60%
12.5	100%

Note that for this purpose, the wall area includes the area of the windows and doors.

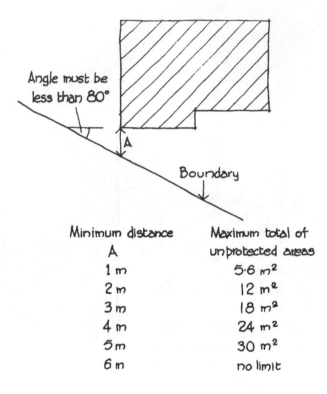

Angle must be less than 80°

A

Boundary

Minimum distance A	Maximum total of unprotected areas
1 m	5·6 m²
2 m	12 m²
3 m	18 m²
4 m	24 m²
5 m	30 m²
6 m	no limit

Fig. 8.3 Permitted unprotected areas in small residential buildings

Instead of the limits given above, the maximum size of windows and doors in the side of a building facing a boundary can be found by reference to any of the other methods set out in BRE Report: *External fire spread: Building separation and boundary distances* (BRE 1991).

8.7 Fire doors

The important property of a fire resisting door is its integrity, i.e. its ability to withstand a fire breaking through the door or its frame. The tests for such a fire door assembly are contained in BS 476: Part 22 and are classified by the number of minutes' resistance; for instance, a door described as FD30 would prevent flame breaking through for 30 minutes. If the suffix S is added, i.e. FD30S, the door also restricts the leakage of smoke, which can be as much of a hazard as the fire itself.

Table 8.6 gives the performance requirements for fire doors in specific locations in terms of their integrity.

Table 8.6 Provisions for fire doors

Position of door	Minimum fire resistance in terms of integrity (minutes)
In a compartment wall separating buildings	As for the wall in which it is fitted but not less than 60
In a compartment wall:	
between a flat or maisonette and a common area enclosing a protected shaft forming a stairway to flats or maisonettes	FD30S FD30S
not described above	As for the wall in which it is fitted*
Forming part of the enclosures of:	
a protected stairway except in a single family house	FD30S
a lift shaft or service shaft	FD30
Forming part of the enclosures of:	
a protected lobby or protected corridor to a stairway	FD30S
a protected corridor other than to a stairway	FD20S
Affording access to an external escape route	FD30
Subdividing:	
corridors connecting alternative exits	FD20S
dead end parts of corridors from the rest of the corridors	FD20S
a dwellinghouse and a garage	FD30
Forming part of the enclosures to:	
a protected stairway in a single family house	FD20
a protected entrance hall or protected landing in a flat or maisonette	FD20
In any other fire resisting construction not described above	FD20

(*Source:* Based on Table B1 of Approved Document B)

* If the door is used for escape the suffix S should be added.

8.7.1 Self-closing fittings

In addition to the door's performance, all fire doors, except cupboard and duct doors which are normally kept shut and locked, should be fitted with an automatic self-closing device.

Where a self-closing device would be considered a hindrance to the occupants of the building in the course of their normal use of the premises, fire doors may be held open by one of the following:

1 a fusible link, unless the doorway is a means of escape;

2 an automatic release mechanism if the door can also be closed manually and it does not lead to either an escape stairway or a fire fighting stair;
3 a door closure delay device.

With the exception of doors in dwellinghouses, those to and within flats or maisonettes and lift doors, fire doors should be marked on both sides with an appropriate fire safety sign complying with BS 5499: Part 1 according to whether the door is:

- to be kept locked when not in use;
- to be kept closed when not in use;
- held open by an automatic release mechanism.

Cupboard and duct doors require fire safety signs on the outside face only.

8.7.2 Ironmongery

The hinges and hardware of a fire door will affect its performance and need to be chosen with care. Guidance is available in a booklet published by the Association of Builders' Hardware Manufacturers in 1983 with the title, *Code of practice for hardware essential to the optimum performance of fire resisting timber doorsets*. Hinges, unless shown to be satisfactory when tested as part of the fire door assembly, should be made from materials with a melting point above 800°C.

Table 8.7 Limitations on uninsulated glazing on escape routes

Position of glazed element	Maximum total glazed area in parts of a building with access to a single stairway	
	Walls	Door leaf
Within a protected stairway in a single family house	Fixed fanlight only	Unlimited
Within a protected entrance hall or landing of a flat or maisonette	Fixed fanlight only	Unlimited above 1.1 m from the floor
Between a protected stairway and the accommodation or unprotected corridor	Nil	25% of the door area
Between the accommodation and a protected lobby or protected corridor forming a dead end	Unlimited above 1.1 m from the floor	Unlimited above 0.1 m from the floor
Between a protected stairway and a protected lobby	Unlimited above 1.1 m from the floor	Unlimited above 0.1 m from the floor

(*Source:* Based on Table A4 of Approved Document B)

8.7.3   Glazed door panels

Because they reduce the fire resistance, uninsulated glazed panels in a door are limited in size as shown in Table 8.7 (opposite).

Chapter 9

STAIRCASES

9.1 The Building Regulations applicable

B1 Means of escape
K1 Stairs and ramps
K2 Protection from falling
M2 Access and facilities for disabled people

Regulation B1 requires that the building be designed and constructed so that there are means of escape from the building in case of fire, to a place of safety outside the building, capable of being safely and effectively used at all material times; K1 states that stairs, ladders and ramps must offer safety to users moving between levels of the building; K2 is concerned that stairs, ramps, floors and balconies, and any roof to which people normally have access, be guarded with barriers where they are necessary to protect users from the risk of falling; and M2 requires that reasonable provision shall be made for disabled people to gain access to and use the building.

The means of escape requirements apply to flats and maisonettes and are principally concerned with the provision of a fire protecting shaft. The whole subject of means of escape is dealt with in Chapter 15.

The requirements of Regulation M2 do not in fact apply to houses, flats or maisonettes, only to public buildings and the like. None the less, the rules laid down set a reasonable standard for access by the disabled and could well be followed in domestic work where the occupants are likely to find moving in and around the building a difficulty. **D9.1**

9.2 The application of Regulations K1 and K2

In dwellings, the requirements of the Regulations apply at any position where there is a difference in floor level in excess of 600 mm.

The requirement does not apply to means of access outside a building unless the access is part of the building; for example, the requirement does not apply to steps in the length of paths leading to the entrance doors but it does apply to any steps at the entrance doors provided that they rise more than 600 mm.

9.3   Definitions

To avoid any ambiguity, the Approved Document K1 sets out the following meaning to terms used in connection with staircases:

Alternating tread stair: a stair of paddle shaped treads with the wide portion alternating from one side to the other on consecutive steps. (This type of stair is more common in loft conversion work and is detailed in Chapter 14.)

Containment: a barrier that prevents people falling from one floor to the storey below.

Flight: the part of a stair or ramp between landings that has a continuous series of steps or a continuous slope.

Going: the horizontal dimension from front to back of a tread less any overlap with the next tread above.

Helical stair: a stair that describes a helix round a central void.

Ladder: a means of access to another level formed by a series of rungs or narrow treads on which a person normally ascends or descends facing the ladder. (Fixed ladders are only to be used in loft conversions and are also detailed in Chapter 14.)

Ramp: a slope of over 1 in 20 designed to take a person from one level to another.

Rise: the height between consecutive treads.

Spiral stair: a stair that describes a helix round a central column.

Tapered tread: a step in which the nosing is not parallel to the nosing of the step or landing above.

Note, particularly, the distinction between a helical stair and a spiral stair.

9.4   Straight flights

It is important that all the steps in a flight of stairs and in each flight where the stair provides access to several floors have exactly the same rise and going and all treads must be level.

In a domestic stair the following rules apply: **D9.2**

- Maximum rise 220 mm.
- Minimum going 220 mm.
- Twice the rise plus the going – between 550 mm and 700 mm.
- Maximum pitch angle 42°.
- Minimum headroom 2.0 m above the pitch line.

- Minimum width – no recommendation given. **D9.3**
- Handrail – at one side at least, between 900 and 1000 mm above the pitch line. **D9.4**
- Guarding – 900 mm above pitch line to withstand a horizontal force of 0.36 kN/m.

In practical terms, the available combinations of rise and going are:

- Any rise between 165 and 200 mm can be used with any going between 223 and 300 mm.
- Any rise between 155 and 220 mm can be used with any going between 245 and 260 mm.

Whether or not the steps have an open riser or closed riser, the setting out is the same but, additionally, each tread in an open riser stair must overlap the one below by at least 16 mm and some provision must be made in such a stair, likely to be used by children under 5 years old, which will prevent the passage of a 100 mm sphere through the open riser. Figure 9.1 shows the setting out of a straight flight.

The headroom dimension must be maintained across the full width of the flight and be unobstructed. Handrails should be provided to both sides of any stairway wider than 1 m. Where the heights match, the handrail can be the top member of the guarding containment. The containment or guarding to a stair likely to be used by children under 5 years old must be designed so that a sphere of 100 mm cannot be passed through, to prevent a child getting caught in it, and must not be easy to climb. **D9.5**

9.5	Landings

Landings must be provided at the top and the bottom of every flight and may include part of the floor of the building. The width and length of each landing must be at least as great as the smallest width of the flight. In addition, if a stair contains more than 36 risers between floors, it must be divided into flights by landings and must change direction by at least 30°.

To afford safe passage, the top landing must be completely clear of any obstructions, the bottom landing and any intermediate landings must be clear of any permanent obstructions but a door may swing across a bottom landing and a cupboard or duct door may swing across an intermediate landing. In both cases, this is only permitted if the door, or doors, leave a clear space of at least 400 mm across the full width of the flight (see Fig. 9.2).

Guarding must be provided to the edge of the landing, 900 mm high, capable of withstanding a horizontal force of 0.36 KN/m.

Landings must be level, unless they are formed by the ground at the top or the bottom of the flight in which case a slope is permitted but it must not exceed 1 in 20.

The minimum headroom height over a landing is 2 m as shown on Fig. 9.1.

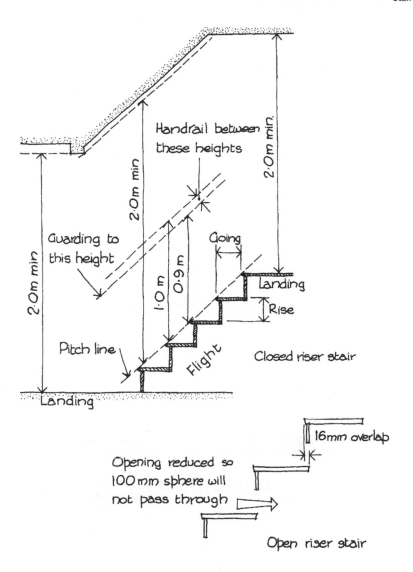

Fig. 9.1 Setting out straight flight stairs

9.6 Tapered treads

A tapered tread has a going which varies across its width. Approved Document K1 rules that, in this case, the going shall be measured at the middle if the step is less than 1 m wide or at 270 mm in from each side if it is over 1 m wide. Where the tread or treads are fitted into a rectangular enclosure, the width for this purpose is taken as being from the inner edge of the flight to a curved line connecting the outer edges of steps above and below (see Fig. 9.3).

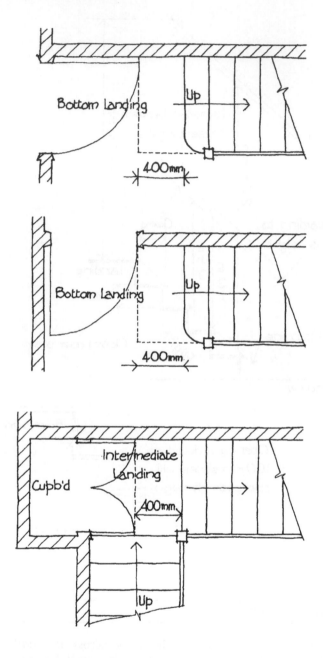

Fig. 9.2 Doors opening over landings

The maximum and minimum going and rise and the rule that twice the rise plus the going should be between 550 and 700 mm applies to the centre line measurement where the stair is less than 1 m wide and, where it is more, the going at the inner measurement line must be not less than the minimum given

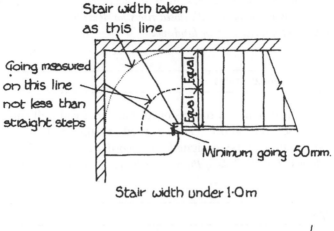

Stair width taken as this line

Going measured on this line not less than straight steps

Equal | Equal

Minimum going 50mm.

Stair width under 1·0m

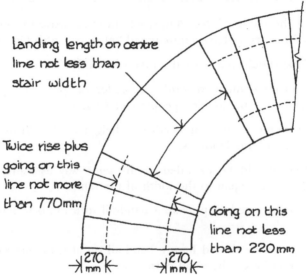

Landing length on centre line not less than stair width

Twice rise plus going on this line not more than 770mm

Going on this line not less than 220mm

270 mm

270 mm

Stair width more than 1·0m

Fig. 9.3 Tapered treads

and that at the outer line not more than the maximum. The going at the inner edge of any tapered tread must be not less than 50 mm.

If there are a number of consecutive tapered treads the going must be uniform for each one and if there are straight treads as well as tapered ones, the going of the tapered treads must be not less than that of the straight treads.

Landings may be provided in the length of a series of tapered treads. Their width is measured in the same way as the treads and their length is measured along the centre line no matter what the width (see Fig. 9.3).

Alternatively, the stair can be designed in accordance with BS 585: *Wood stairs*, Part 1: *Specification for stairs with closed risers for domestic use, including straight and winder flights and quarter or half landings.*

9.7 Spiral and helical stairs

As any spiral or helical stair is a succession of tapered treads, all the rules and recommendations set out above can be used as the basis for a design. Alternatively, the stair can be designed in accordance with BS 5395: *Stairs, ladders and walkways*, Part 2: *Code of Practice for the design of helical and spiral stairs.*

9.8 Ramps

In many respects the standards for ramps, set out in Approved Document K1, are the same as for stairs and are summarized below (see also Fig. 9.4): **D9.6**

Steepness: the slope should not exceed a rake of 1 in 20.

Width: there is no recommended width except for ramps forming means of escape and those for use by disabled people (see Chapter 15).

Headroom: there should be a clear headroom height of 2 m throughout the length of the ramp and any landings.

Landings: landings should be provided in the same way as stairs and their length should be at least equal to the width of the ramp.

Handrails: there is no need to provide a handrail to a ramp which rises less than 600 mm.

Ramps under 1 m wide should have a handrail on at least one side, ramps over 1 m wide should have a handrail both sides.

The height of the handrail should be between 900 and 1000 mm. They should give a firm support and good grip and can form the top of the containment guarding if the heights match.

Guarding: ramps and landings in domestic work must be guarded by a barrier 900 mm high which can withstand a horizontal force of 0.36 kN/m and, as for stairs, it should neither be possible to pass a 100 mm sphere through it, nor should the barrier be easy to climb.

Obstructions: ramps must be clear of all obstructions.

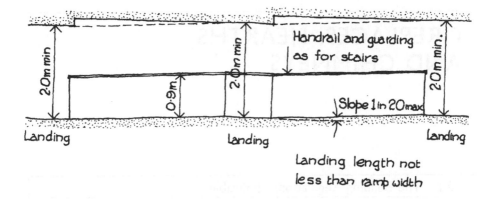

Fig. 9.4 Setting out ramps

See D9.7 in Appendix D on the subject of lifts.

Chapter 10

FIREPLACES, HEARTHS AND CHIMNEYS

10.1 The Building Regulations applicable

A1 Loading
B3 Internal fire spread (structure)
J1 Air supply
J2 Discharge of products of combustion
J3 Protection of building

A1 concerns the loads imposed on the building and its parts and requires that they shall be so constructed that they can safely withstand the combined wind, dead and imposed loads, chimneys are subjected to wind loads and so are covered by this regulation; B3 refers, very briefly, to flues in connection with compartment walls and requires that the construction shall inhibit the spread of fire; J1 is concerned that there is an adequate supply of air to an appliance for combustion and for the efficient working of any flue pipe or chimney; J2 states that heat producing appliances should have adequate provision for the discharge of the products of combustion to the outside air; and J3 requires that heat producing appliances and flue pipes should be so installed, and fireplaces and chimneys so constructed, as to reduce to a reasonable level the risk of the building catching fire in consequence of their use.

The following recommendations are based on Approved Document A1/2 Section 1D and J1/2/3 and apply to solid fuel appliances. Gas fired appliances with a rated input up to 60 kW and oil fired appliances with a rated output up to 45 kW are dealt with in Chapter 11.

10.2 Fire recesses and hearths

Fireplace recesses are only required for some solid fuel burning appliances designed to burn within such an enclosure. Other types of solid fuel fire or, more particularly, boiler, are designed specifically to stand out in the room on a hearth.

10.2.1 Fireplaces for solid fuel appliances

The Regulations are solely concerned with the structure surrounding the fire, i.e. the fireplace recess and a constructional hearth both of which must be formed of solid non-combustible material.

The brick or blockwork each side of the recess must be not less than 200 mm thick. At the back of the recess, the thickness should be 200 mm if there is a room the other side of the wall or 100 mm if there is another fireplace recess back-to-back with it. If the recess is in an external wall the back should be at least 100 mm thick if the external wall is solid but if it is a cavity wall with the cavity running through, each leaf must be not less than 100 mm (see Fig. 10.1).

Hearths must be at least 125 mm thick, should project from the face of the chimney breast by at least 500 mm and overlap the brickwork each side of the recess by at least 150 mm. The thickness may include the thickness of any non-combustible floor under the hearth. Combustible material under a constructional hearth should be either separated from the underside of it by an air space of at least 50 mm, or not less than 250 mm below the top surface of the hearth. The only exception to this rule is combustible material which is allowed as support around the edges of the hearth (see Fig. 10.1).

10.2.2 Hearths for free-standing solid fuel appliances

Any free-standing solid fuel appliance must be placed on a constructional hearth which conforms to the same standards as a hearth in a recess. In plan, the hearth should project 150 mm each side and behind the fitting and 225 mm in front, if it is a closed stove, or 300 mm in front, if it is a type which, it is intended, can be used with the front open. If the hearth is larger than the minimum that this would give, combustible material may be placed on it but no nearer than 150 mm each side and behind the appliance and 225 mm or 300 mm in front.

Any wall behind or beside the appliance and less than 150 mm from the edge of the hearth should be of non-combustible material for a height of at least 1.2 m and for a thickness of at least 75 mm. If the wall is between 150 and 50 mm from the appliance, the non-combustible material must extend at least 300 mm above the fitting and, if it is under 50 mm from the appliance, its thickness must be increased to 200 mm (see Fig. 10.2).

10.3   Flues and flue pipes

Flue pipes for solid fuel appliances may be made in any of the materials listed below but should only be used to connect the appliance to a chimney. In addition they should not pass through a roof space.

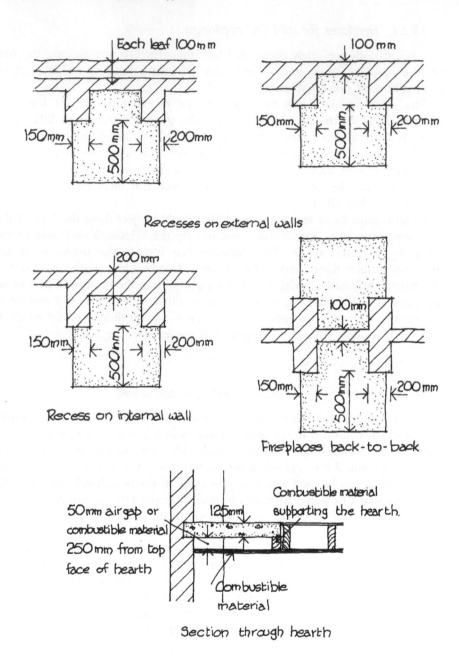

Each leaf 100mm

150mm 500mm 200mm

100 mm

150mm 500mm 200mm

Recesses on external walls

200mm

150mm 500mm 200mm

Recess on internal wall

100mm

150mm 500mm 200mm

Fireplaces back-to-back

50mm air gap or
combustible material
250mm from top
face of hearth

125mm

Combustible material
supporting the hearth.

Combustible
material

Section through hearth

Fig. 10.1 Fireplace recesses for solid fuel fires

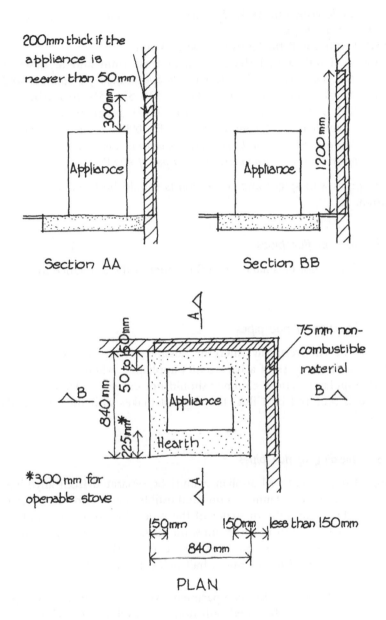

200mm thick if the appliance is nearer than 50mm

300mm

Appliance

Section AA

1200 mm

Appliance

Section BB

A

B

50 to 150mm

840 mm

225mm*

Appliance

Hearth

75mm non-combustible material

B

A

*300 mm for openable stove

150mm   150mm   less than 150mm

840 mm

PLAN

Fig. 10.2 Hearths for free-standing solid fuel appliances

149

1 Cast iron as described in BS 41: *Specification for cast iron spigot and socket flue or smoke pipes and fittings.*
2 Mild steel with a wall thickness of at least 3 mm.
3 Stainless steel with a wall thickness of at least 1mm and as described in BS 149: *Steel plate, sheet and strip,* Part 2: *Specification for stainless and heat resisting steel plate, sheet and strip* for Grade 316 S11, 316 S13, 316 S16, 316 S31, 316 S33, or the equivalent Euronorm 88–71 designation.
4 Vitreous enamelled steel complying with BS 6999: *Specification for vitreous enamelled low carbon steel flue pipes, other components and accessories for solid fuel burning appliances with a maximum rated output of 45 kW.*

All flue pipes with spigot and socket joints should be fitted with the socket facing upwards.

10.3.1 Sizes of flue pipes

A flue pipe should not be smaller than the outlet of the appliance to which it is connected.

10.3.2 Direction of flue pipes

Flues should be run vertically wherever possible. Horizontal runs should be avoided except in the case of a back outlet appliance, when the length of flue between the appliance and a chimney should not exceed 150 mm.

If a bend is required in a flue, the angle it makes with the vertical should not exceed 30°.

10.3.3 Shielding of flue pipes

A flue pipe from a solid fuel appliance must be separated from anything that might burn by at least 200 mm of non-combustible material or an air space of not less than three times the diameter of the pipe. If a non-combustible shield is provided, equal in width to three times the diameter of the pipe and spaced away from the combustible material by at least 12.5 mm, the air space required can be reduced to one and a half times the diameter of the pipe (see Fig. 10.3).

If the wall or floor is of the compartment type as described in Chapters 5 and 7, the flue pipe must be cased with non-combustible material with at least half the fire resistance specified for the wall or floor.

10.3.4 Outlets from flues

The outlet from a flue, contained within a chimney, serving a solid fuel appliance must be carefully positioned in relation to the roof line as shown in Figs 10.4 and 10.5.

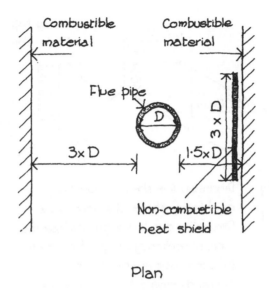

Plan

Fig. 10.3 Separation of solid fuel flue pipe and combustible material

As can be seen from Fig. 10.4 the Approved Document does not give a direct height of chimney above a pitched roof, instead it gives a zone outside which the flue outlet must be placed. The effect of this is that the actual minimum height of chimney required is not fixed, except in relation to the ridge, and is determined by the pitch of the roof. The graph in Fig. 10.4 is an interpretation of the Approved Document rule and from it can be read both the direct height and the distance from the ridge to the point where the required height becomes 600 mm above the ridge.

For the frequently used pitch of 35° the minimum height of the flue outlet above the highest point of intersection of the chimney with the roof is 1615 mm but if it is within 1443 mm of the ridge the top should be 600 mm above the ridge.

The outlet from a flue attached to a pressure jet oil burning appliance can be terminated anywhere above the roof.

10.4   Chimneys

Chimneys can be built to serve solid fuel, gas fired or oil fired appliances and, to a very large extent their construction is the same no matter what the fuel.

10.4.1   Brick and block chimneys

Brick chimneys serving any type of appliance should be lined. The liners should be of one of the types listed below, fitted with the socket or rebate

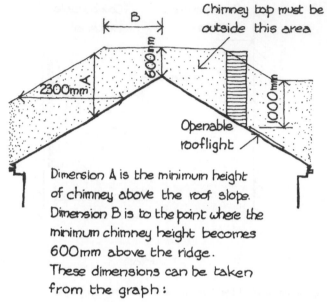

Dimension A is the minimum height of chimney above the roof slope. Dimension B is to the point where the minimum chimney height becomes 600mm above the ridge.
These dimensions can be taken from the graph:

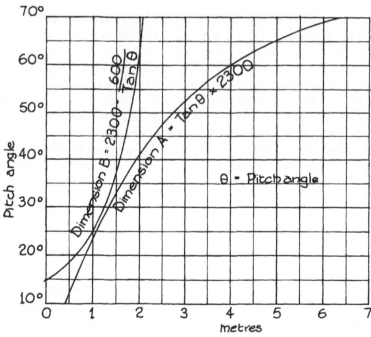

Fig. 10.4 Height of chimney above a pitched roof

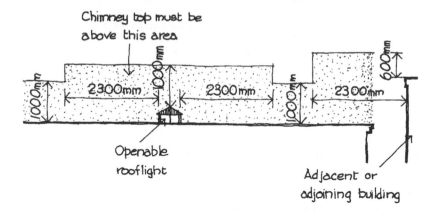

Fig. 10.5 Height of chimney above a flat roof

facing upwards to prevent any condensate from running out, and jointed with fire-proof mortar. Any space between the liners should be filled with weak mortar or insulating concrete.

Types of liner listed in Approved Document J1/2/3 are:

1 Clay flue liners with rebated or socket joints as described in BS 1181: *Specification for clay flue linings and flue terminals.*
2 Imperforate clay pipes with socketed joints as described in BS 65: *Specification for vitrified clay pipes, fittings and joints.*
3 High alumina cement and kiln burnt or pumice aggregate pipes with rebated or socketed joints or steel collars around the joints.

A blockwork chimney serving a solid fuel appliance should be either lined as for a brick chimney, or built from refractory materials or a combination of high alumina cement and kiln-burnt or pumice aggregate. Blockwork chimneys for gas fired appliances and oil fired appliances with flue gases cooler than 260°C can be built with purpose-made flue blocks conforming either to BS 1289: *Flue blocks and masonry terminals for gas appliances*, Part 1: *Specification for precast concrete flue blocks and terminals* or Part 2: *Specification for clay flue blocks and terminals.*

The size of the flue inside a chimney should never be less than the size of the outlet of the appliance connected to the chimney, or the size recommended by the manufacturer of the appliance, or as shown in Table 10.1.

The walls of a chimney, excluding the liners, should be not less than 100 mm thick unless the wall is between the flue and either another compartment of the same building, another building or another dwelling, in which case it must be 200 mm thick.

Table 10.1   Sizes of flues in chimneys serving solid fuel appliances

Installation	Minimum circular flue size (or square section of equivalent area)
Fireplace recess with an opening up to 500 mm × 550 mm	200 mm
Fireplace recess with an opening more than 500 mm × 550 mm	A free area of 15% of the area of the recess opening
Closed appliance up to 20 kW rated output burning bituminous coal	150 mm
Closed appliance up to 20 kW rated output	125 mm
Closed appliance from 20 to 30 kW rated output	150 mm
Closed appliance from 30 to 45 kW rated output	175 mm

(*Source:* Based on Table 2 of Approved Document J1/2/3)

Combustible material should be separated from the chimney by either 200 mm from the flue or 40 mm from the outer face of the chimney or fireplace recess, unless it is a floorboard, skirting, dado or picture rail, mantleshelf or architrave. Metal fixings in contact with combustible materials should be at least 50 mm from the flue.

10.4.2   Proportions of brick and block chimneys

Unless the chimney is adequately supported by ties or securely restrained by any other means, the height should not exceed 4.5 times the smallest width. The height is measured from the highest point of intersection with the roof surface and the width is measured at this same level. Any chimney pot or flue terminal is included in the height, see Fig 10.6.

10.4.3 Factory-made chimneys

Factory-made insulated chimneys should be constructed and tested to meet the relevant recommendations given in BS 4543 *Factory-made insulated chimneys*, Part 1: *Methods of test for factory-made insulated chimneys*, Part 2: *Specification for chimneys for solid fuel appliances*, or Part 3: *Specification for chimneys with stainless steel flue linings for use with oil fired appliances*, as appropriate. They should be installed in accordance with the manufacturer's instructions or to meet the relevant recommendations of BS 6461: *Installation of chimneys and flues for domestic appliances burning solid fuel (including wood and peat)*, Part 2: *Code of practice for factory-made insulated chimneys for internal applications*. Although BS 6461 relates

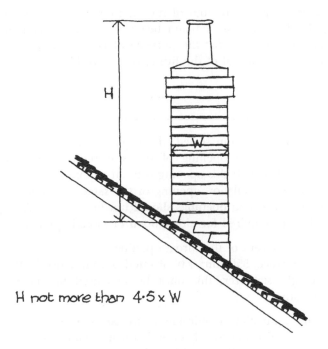

H not more than 4·5 x W

H - height from highest point of roof
   intersection to top of chimney pot
W - least horizontal dimension

Fig. 10.6 Proportions of a masonry chimney

specifically to solid fuel appliances, its recommendations should also be followed if the chimney is to serve either a gas fired or oil fired appliance.

A factory-made insulated chimney or an insulated metal chimney intended to serve a gas or oil fired appliance, should not pass through any part of the building unless it is cased in non-combustible material giving at least half the fire resistance of the compartment wall or floor. Nor should it pass through a cupboard, storage space or roof unless it is surrounded by a non-combustible guard fixed at a distance from the outer surface of the chimney shown to be safe by test in accordance with BS 4543: Part 1. This same safe distance applies to the spacing which must be allowed between the outer wall of a factory-made insulated chimney and any combustible material.

10.4.4   Debris collecting space

Any chimney to a solid fuel appliance which is not directly over the fitting or one serving a gas or oil fired appliance which is not lined or not constructed of purpose-made flue blocks must be provided with a debris collecting space.

This space must be at the bottom of the chimney, have a volume of at least $0.012 \, \text{m}^3$, a depth of at least 250 mm below the point of connection of the appliance and be accessible for the purpose of clearing any debris collected. Access requiring the removal of the fitting is permissible.

10.5 Air supply

Unless the appliance is room sealed, the room or space in which it is sited should have a ventilation opening of the size required for the type and capacity of the appliance. If this opening is to an adjoining room or space then that too must have a ventilation opening of the same size which vents directly to the open air. Ventilation openings should not be in fire resisting walls.

The specific sizes required for either open or closed appliances are:

- Open fire, 50 per cent of the throat opening.
- Any other appliance, $550 \, \text{mm}^2/\text{kW}$ of rated output above 5 kW.
  With a draught stabilizer, this must be increased by $300 \, \text{mm}^2$ for each kilowatt of output.

Because of the ventilation requirements to be observed in a kitchen, no internal kitchen (which has to rely entirely on mechanical ventilation) should contain an open flued appliance as there is a possiblility of flue gas spillage.

Natural ventilation only should be provided to any kitchen containing a solid fuel appliance.

10.6 Alternative approaches

The requirements of Regulations J1, J2 and J3 may also be met by compliance with the relevant sections of the following British Standard:

- BS 8303: *Code of practice for installation of domestic heating and cooking appliances burning solid mineral fuels.*

Chapter 11

GAS AND OIL FIRED EQUIPMENT, HEATING AND HOT WATER

11.1 The Building Regulations applicable

G3   Hot water storage
J1   Air supply
J2   Discharge of products of combustion
J3   Protection of buildings
L1   Conservation of fuel and power

G3 requires that unvented hot water systems are safe; J1 sets standards for an adequate supply of air for combustion purposes; J2 states that heat producing appliances must have adequate provision for the discharge of the products of combustion to the outside air; J3 requires that heat producing appliances and flue pipes be so installed as to reduce to a reasonable level the risk of the building catching fire in consequence of their use; and L1 includes as one of the reasonable provisions for the conservation of fuel and power, the controlling of the operation of space heating and hot water systems and limiting the heat loss from hot water cylinders, hot water pipes, heating pipes and hot air ducts.

The following recommendations are based on Approved Documents G3, J1/2/3 and L1 and refer to gas and oil fired appliances. Solid fuel burning equipment is dealt with in Chapter 10.

11.2 Hearths

Gas and oil fired appliances are all contained within themselves (except for gas fired fittings which simulate the burning of wood or coal) and do not need to be set in a fire recess. Approved Document J1/2/3 sets out recommendations for hearths suitable to receive these fires and boilers.

Gas fires which simulate the burning of wood or coal generally do require a fire recess and should be installed in accordance either with the manufacturer's recommendations, where these have been tested and passed by an approved authority, or as given for solid fuel appliances in Chapter 10, Section 10.2.1. Alternatively, the recommendations of BS 5871: *Specification for*

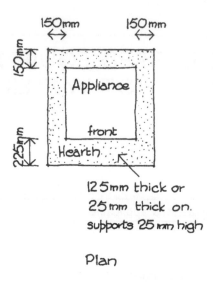

Fig. 11.1 Hearths for back boiler gas appliances

installation of gas fires, convector heaters, fire/back boilers and decorative fuel effect gas appliances should be followed.

11.2.1 Hearths for gas fired appliances

The following does not apply to gas fires which simulate the burning of wood or coal, for this type of appliance provisions should be made as set out above.

A hearth should always be provided under a gas fired appliance unless every part of the flame or incandescent material is at least 225 mm above the floor or the appliance complies with the recommendations of BS 5258: *Safety of domestic gas appliances* or BS 5386: *Specification for gas burning appliances for installation without a hearth.*

In the case of a fitting which includes a back boiler and requires a hearth, its dimensions should be 150 mm larger than the appliance at the sides and back, 225 mm larger in front and either 125 mm thick and solid or 25 mm thick on non-combustible supports 25 mm high (see Fig. 11.1).

Any other appliance which does not include a back boiler should be installed on a non-combustible hearth of the same dimensions but which only needs to be 12 mm thick.

11.2.2 Hearths for oil fired appliances

If the surface temperature of the floor below the appliance is likely to exceed 100°C, then a constructional hearth is required. In plan, the hearth should be the same as for a gas appliance, i.e. it should project 150 mm each side and behind the fitting, 225 mm in front and be 125 mm thick. If the hearth is larger than the minimum that this would give, combustible material may be placed on it but no nearer than 150 mm each side and behind the appliance and 225 mm in front.

Oil fired appliances which are not likely to raise the temperature of the floor surface above 100°C may stand on a rigid, imperforate sheet of non-combustible material without the need for a constructional hearth.

11.3   Flues and flue pipes

Flue pipes for any oil fired appliance in which the flue gas temperature is likely to exceed 260°C may be made in any of the materials listed below but should only be used to connect the appliance to a chimney. In addition, they should not pass through a roof space.

1 Cast iron as described in BS 41: *Specification for cast iron spigot and socket flue or smoke pipes and fittings.*
2 Mild steel with a wall thickness of at least 3 mm.
3 Stainless steel with a wall thickness of at least 1mm and as described in BS 149: *Steel plate, sheet and strip*, Part 2: *Specification for stainless and heat resisting steel plate, sheet and strip* for Grade 316 S11, 316 S13, 316 S16, 316 S31, 316 S33, or the equivalent Euronorm 88–71 designation.
4 Vitreous enamelled steel complying with BS 6999: *Specification for vitreous enamelled low carbon steel flue pipes, other components and accessories for solid fuel burning appliances with a maximum rated output of 15 kW.*

Flue pipes for all gas fired appliances and any oil fired appliances in which the temperature of the flue gas is not likely to exceed 260°C may be made in any of the following:

1 Any of the materials listed above.
2 Sheet metal as described in BS 715: *Specification for metal flue pipes, fittings, terminals and accessories for gas fired appliances with a rated input not exceeding 60 kW.*
3 Asbestos cement as described in BS 567: *Specification for asbestos-cement flue pipes and fittings, light quality*, or BS 835: *Specification for asbestos-cement flue pipes and fittings, heavy quality.*
4 Cast iron as described in BS 41.
5 Any other material fit for its intended purpose.

All flue pipes with spigot and socket joints should be fitted with the socket facing upwards.

A flexible flue liner may be used in a chimney venting a gas appliance or an oil fired appliance in which the flue gas temperature is not likely to exceed 260°C, provided that the liner complies with the requirements of BS 715 and the chimney either was built before 1 February 1966 or is already lined or constructed of flue blocks as described in Chapter 10.

11.3.1 Sizes of flue pipes

Flue pipes to balanced flue units, solid fuel simulating gas fires and oil burning fittings designed for low level flues should be sized in strict accordance with the recommendations of the manufacturer.

A circular flue pipe fitted to any other type of gas fire should have a cross-sectional area of not less than $12\,000\,\text{mm}^2$ and a rectangular pipe similarly fitted should have a cross-sectional area of not less than $16\,500\,\text{mm}^2$ with a minimum dimension in either direction of 90 mm.

Apart from any special provisions specified in appliances described above, whatever fuel is used, the minimum size for a flue pipe should be not less than the size of the flue outlet on the appliance.

11.3.2 Direction of flue pipes

Flues should be run vertically wherever possible. Horizontal runs should be avoided except in the cases of a back outlet appliance, when the length of flue between the appliance and a chimney should not exceed 150 mm, or a balanced flue appliance installed as recommended by the manufacturer.

If a bend is required in a flue, the angle it makes with the vertical should not exceed 45°.

11.3.3 Shielding of flue pipes

Flue pipes to gas fired appliances and oil fired appliances which have flue gas temperatures less than 260°C should be positioned at least 25 mm away from any combustible material. Where they pass through a wall, floor or roof, they should be separated from anything that can burn by a non-combustible sleeve which encloses an air space of not less than 25 mm all round the flue pipe. For a double walled flue pipe, the 25 mm can be measured from the outside of the inner pipe.

If the wall or floor is of the compartment type as described in Chapters 5 and 7, the flue pipe must be cased with non-combustible material with at least half the fire resistance specified for the wall or floor.

11.3.4 Outlets from flues

The outlet from a flue serving an oil fired appliance (other than one of the pressure jet or balanced flue type) must be positioned above the roof line in the same manner as that required for a solid fuel appliance and as shown on Figs 10.4 and 10.5 of the previous chapter.

Gas fired appliances designed to operate with a balanced flue should have their outlet positioned so that there is a free intake of air and dispersal of the products of combustion. The outlet should be at least 300 mm from any opening wholly or partly above the terminal which should be fitted with a guard if it is likely to be touched by a person or damaged in any way. It should also be designed to prevent the entry of anything that might affect the operation of the flue.

The outlet from any other gas fired appliance must so be sited, at roof level, that air may pass freely across it at all times and be 600 mm from any opening into the building. If the flue, except that to a gas fire, is more than 175 mm across in any direction it should be fitted with a flue terminal.

A balanced flue or a low level discharge flue from an oil fired appliance can terminate in the same way as a balanced flue from a gas fired appliance except that it should be not less than 600 mm from any opening into the building.

11.4   Chimneys

As the construction of a chimney to remove the products of combustion from a gas or oil fired appliance is principally the same as that serving a solid fuel burning appliance the whole subject has been dealt with in the previous chapter under Section 10.4 to which reference should be made.

11.5   Shielding of appliances

Gas appliances that do not comply with the relevant parts of BS 5258 or BS 5368 and oil fired appliances with a surface temperature of the sides or back in excess of 100°C need to be separated from anything that might burn by a shield of non-combustible material at least 25 mm thick or an air space of not less than 75 mm.

11.6   Air supply

Unless the appliance is room sealed, the room or space in which it is sited should have a ventilation opening of the size required for the type and capacity of the appliance. If this opening is to an adjoining room or space then that too must have a ventilation opening of the same size which vents directly to the open air. Ventilation openings should not be in fire resisting walls.

The specific size required for each type of appliance is:

- Gas cooker: an openable window or other means of ventilation of any size. If the room volume is less than $10 \, m^3$ a permanent vent of $5000 \, mm^2$ is needed.
- Open flued gas appliance: $450 \, mm^2$ for each kilowatt of input over 7 kW.

- Any oil fired appliance: 550 mm^2 for each kilowatt of output over 5 kW.

Because of its ventilation requirements, no internal kitchen (which has to rely solely on mechanical ventilation) should contain an open flued appliance as there is a risk of flue gas spillage.

11.7 Alternative approaches

The requirements of Regulations J1, J2 and J3 may also be met by compliance with the relevant sections of the following British Standards:

- BS 5546: *Specification for installation of gas hot water supplies for domestic purposes (2nd family gases).*
- BS 5864: *Specification for installation of gas-fired ducted-air heaters of rated input not exceeding 60 kW (2nd family gases).*
- BS 5440: *Installation of flues and ventilation for gas appliances of rated input not exceeding 60 kW (1st, 2nd and 3rd family gases,* Parts 1 and 2.
- BS 5871: *Installation of gas fires, convector heaters, fire/back boilers and decorative fuel effect gas appliances,* Parts 1, 2 and 3.
- BS 6172: *Specification for installation of domestic gas cooking appliances (2nd family gases).*
- BS 6798: *Specification for installation of gas fired hot water boilers of rated input not exceeding 60 kW.*
- BS 5410: *Code of practice for oil firing,* Part 1: *Installations up to 44 kW output for space heating and hot water supply purposes.*

11.8 Appliances in bathrooms and garages

As required by the Gas Safety (Installation and Use) Regulations 1984, any appliance fitted in a bathroom or garage must be of the room sealed type.

11.9 Unvented hot water systems

The requirements of Regulation G3 do not apply to a hot water storage system that has a storage vessel of 15 litres or less, nor do they apply to a system providing space heating only.

Any system that incorporates a storage vessel that has no vent pipe must be installed by a competent person. It must have safety devices that prevent the temperature of the stored water from exceeding 100°C at any time and it must have pipework that safely conveys the discharge of hot water from the built-in safety devices to a point where it is visible but which will not cause any danger to people in the building.

A competent person is defined in Approved Document G3 as one holding a current Registered Operative card for the installation of unvented domestic hot water systems issued by:

1 the Construction Industry Training Board (CITB);
2 the Institute of Plumbing;
3 the Association of Installers of Unvented Hot Water Systems;
4 an equivalent body.

Also competent are individuals who are designated Registered operatives and employed by companies included on the list of Approved Installers published by the British Board of Agrément up to 31 December 1991.

11.9.1 Design

Any unvented hot water storage system should be in the form of a proprietary package or unit which is:

1 approved by a member body of the European Organization for Technical Approvals (EOTA) operating a technical approvals scheme such as the British Board of Agrément;
2 approved by a certification body accredited by the National Accreditation Council for Certification Bodies (NACCB) which has tested the system against an appropriate standard such as BS 7206: *Specification for unvented hot water storage units and packages*;
3 the subject of a proven independent assessment that will clearly demonstrate an equivalent level of verification and performance.

11.9.2 Direct heating units

Any direct heating unit must have at least two temperature safety devices operating in sequence. One must be a non self-resetting thermal cut-out in accordance with BS 3955: *Specification for electrical controls for household and similar general purposes*, or BS 4201: *Specification for thermostats for gas burning appliances*. The other must be one or more temperature relief valves to BS 6283: *Safety and control devices for use in hot water systems*, Part 2: *Specification for temperature relief valves for pressures from 1 bar to 10 bar*, or Part 3: *Specification for combined temperature and pressure relief valves for pressures from 1 bar to 10 bar*. These devices must be in addition to any thermostats which are fitted to maintain the temperature of the stored water.

Alternatively, the unit could be fitted with other safety devices which are capable of providing an equivalent degree of safety in preventing the temperature of the stored water from exceeding 100°C at any time. These devices must be approved by either a member of EOTA or a body possessing NACCB accreditation or be the subject of a proven independent assessment.

A temperature relief valve should be fitted directly on to the storage vessel in such a way as to ensure that the temperature of the stored water does not

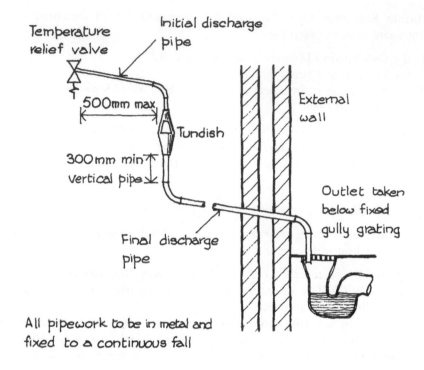

Fig. 11.2 Discharge pipework from an unvented hot water system

exceed 100°C. The valve should be sized so that the discharge rating is not less than equal to the power input to the water. This discharge rating is to be measured in accordance with Appendix F of Part 2 of BS 6283 or Appendix G of Part 3 of the same British Standard.

The discharge from a temperature relief valve must be via a short length of metal pipe, of a size not less than the nominal outlet size of the valve, through an air break over a tundish (see Fig. 11.2). Where several valves are fitted, they may discharge through a manifold which must be sized to accept the total discharge from the pipes connected to it.

11.9.3 Indirect heating units

The safety devices listed for direct heating units are also required for indirect units but the non-self-resetting thermal cut-outs should be wired to a motorized valve or a similar suitable device or they should shut off the flow to the primary heater. If the unit incorporates a boiler, the thermal cut-out may be fitted to it.

The temperature relief valve should be sized and fitted and a discharge pipe provided in the same manner as that described for direct heating units.

Table 11.1 Non-vented heating systems. Sizing of final discharge pipe from temperature relief valve

Valve outlet size	Size of initial discharge pipe (mm)	Size of final discharge pipe (mm)	Maximum resistance allowed expressed as a length of straight pipe (m)	Resistance created by an elbow expressed as a length of pipe (m)
		22	up to 9	0.8
G½	15	28	up to 18	1.0
		35	up to 27	1.4
		28	up to 9	1.0
G¾	22	35	up to 18	1.4
		42	up to 27	1.7
		35	up to 9	1.4
G1	28	42	up to 18	1.7
		54	up to 27	2.3

(*Source:* Based on Table 1 of Approved Document G3)

Indirect heating units, which have an alternative direct method of water heating fitted, need a non-self-resetting thermal cut-out on this alternative direct source as well as on the indirect.

11.9.4 Discharge pipes

The discharge pipe from the vessel as far as and including the tundish is usually a part of the proprietary package. If not, it must be fitted as part of the installation work and, in either case, the tundish must be fixed vertically, in the same space as the water storage system and 500 mm or less from the temperature relief valve.

The pipe from the tundish must be of metal and terminate in a position which will not create a danger to anybody in the vicinity. The minimum diameter of this pipe is one pipe size larger than the nominal outlet size of the temperature relief valve. This assumes a straight pipe of a maximum length of 9 m. If the actual pipe is longer than this or has any elbows or bends, the hydraulic resistance of a 9 m pipe would be exceeded and the diameter would, therefore, need to be increased. Table 11.1 shows the sizing of copper discharge pipes.

The method of sizing is best explained by the following example:

- Check the suitability of a 22 mm copper discharge pipe 7 m long with four elbows serving a $G\frac{1}{2}$ temperature relief valve.
- The maximum resistance allowed for a 22 mm copper pipe from a $G\frac{1}{2}$ valve, expressed as a length of straight pipe, is 9.0 m.
- The reduction required to take account of the resistance of the four elbows is: 4×0.8 m, i.e. 3.2 m.
- The maximum permitted length is $9.0 - 3.2$, i.e. 5.8 m. As this is less than the actual length of 7 m the next larger size of pipe – 28 mm – must be checked and would be found to be satisfactory.

There must be a vertical section of pipe, at least 300 mm long, immediately below the tundish and the rest of the pipework must be fixed with a continuous fall.

It is preferable that the discharge both at the tundish and at the termination of the pipework can be seen, but where this is not possible or presents practical difficulties, one or other of the points of discharge must be clearly visible. Ideally, the final discharge should be to a gully at a point below the grating but above the water seal. Alternatively, it can discharge vertically above an external surface such as a car park, hardstanding or grassed area at a height not exceeding 100 mm above the surface. If there is a possibility that children might come into contact with the end of the discharge pipe a wire cage or similar guard must be fitted which will prevent injury but not restrict the view of the pipe.

High level discharge is also permitted but it must still be visible and if a hopper head and downpipe are provided to receive it, they must be of metal. If the discharge is on to a roof, it must be capable of withstanding the high temperature water, the discharge must be at least 3 m from any plastic guttering which would collect the discharged water and the tundish must also be visible. Note, in this connection, that as the discharge would consist of scalding water or steam, asphalt roofing and bituminous roofing felt would be damaged by it.

A single pipe can serve a number of flats or maisonettes but the maximum should be restricted to six so that any installation discharging can be traced with reasonable ease. The common discharge pipe must be sized in relation to the largest individual discharge pipe connected to it.

In unvented hot water storage systems installed in premises where the safety devices may not be apparent to the occupants – dwellings such as those occupied by blind, infirm or disabled people – consideration should be given to fitting an electronically operated device to warn when a discharge takes place.

11.9.5 Electrical connections

Electrical non-self-resetting thermal cut-outs should be connected to either the direct heat source or the indirect primary flow control device in strict accordance with the current Regulations for Electrical Installations of the Institute of Electrical Engineers.

11.10   Heating controls and insulation of ducts, hot water cylinders and pipes

---

In addition to the requirements about the conservation of fuel and power by the reduction of heat losses through the external fabric of the building, the Regulations also require that provision shall be made to save energy by reducing any waste that could occur in the space heating and hot water systems of the dwelling.

Guidance is given, in Approved Document L, on control systems that would satisfy the requirements of Regulation L1. The guidance given relates to the more common forms of heating system found. It does not deal with individual solid fuel, gas or electric fires or room heaters that, it is assumed will have their own, integral, controls.

11.10.1   Space heating controls

Three provisions must be made to meet the requirements:
- zone controls;
- timing controls;
- boiler control interlocks.

Zone controls aim to economise on energy by applying separate controls to those areas of the dwelling that have differing heating demands, such as the living areas as opposed to the sleeping areas.

Most hot water central heating systems, fan assisted electric storage heaters and electric panel heaters can be operated through room thermostats or thermostatic radiator valves to give independent control of the temperature.

In most dwellings, two zones would be appropriate and adequate to effect an economy but in single storey open plan flats or bedsitters, for example, no great benefit would be obtained and a single zone could be considered.

Ducted warm air systems and flap controlled electric storage heaters do not lend themselves to zone controlling but should be fitted with thermostats.

Timing controls are the central heating programmers now widely used in houses and flats. The Approved Document points out that such timing controls are not suitable for solid fuel boilers that operate only by natural draught.

Boiler control interlocks are designed to stop unnecessary boiler cycling. In many systems, there are two thermostats in addition to any built into the boiler, a room thermostat and a cylinder thermostat. The system should only fire up when either of these thermostats is calling for heat. When no heat is required for either purpose the boiler should be switched off. An alternative to providing room thermostats is to fit thermostatic radiator valves. If these are installed there should also be some form of flow control or other device to prevent the boiler running when it is not required.

11.10.2 Hot water cylinders

It is recommended, in Approved Document L, that the hot water storage vessel should comply with BS 1566 or BS 3198, particularly in relation to the surface areas and pipe diameters of the heat exchanger.

The common size of hot water cylinder is 120 litres (450 mm dia × 900 mm high). This should be insulated with factory applied insulation that restricts standing heat losses to 1 W/l when tested by the method set out in BS 1566: Part 1: Appendix B.4. Cylinders and storage vessels of other sizes should be insulated with the same material and of the same thickness as a 120 litre cylinder. A provision that satisfies the requirements of this Regulation is a factory applied coating of PU foam 35 mm thick. The foam should have zero ozone depletion potential and a minimum density of 30 kg/m.

A thermostat should be fitted to the cylinder that will shut off the supply of heat when the set temperature is reached and, if there is a hot water space heating system, is linked to the room thermostat so that the boiler is shut down when no heat at all is required.

A timer should also be fitted to stop the supply of heat at the times when water heating is not required. This can be, and often is, part of the general central heating programmer but it can also be a local device just for the hot water.

Systems heated by a solid fuel boiler should be fitted with a thermostatic valve unless the cylinder provides the slumber load.

11.10.3 Insulation of pipes and ducts

In some installations, the heat loss from a pipe, or, more rarely, from the surface of a duct is allowed in the calculation of heating surfaces required. If not, then the pipe or duct should be insulated to reduce energy wastage.

Pipe insulation should consist of:

1 a material with a thermal conductivity not greater than 0.045 W/mK (which covers most of the common insulating products);
2 a thickness equal to the outside diameter of the pipe up to a maximum of 40 mm.

Approved Document L1 states that duct insulation (and, alternatively, pipe insulation) should meet the recommendations of BS 5422: *Method for specifying thermal insulating materials on pipes, ductwork and equipment (in the temperature range −40°C to +700°C)*.

The British Standard sets out economic thicknesses for both duct and pipe insulation, taking into account the average cost of the materials and the labour cost of installation against the cost of the heat lost over a five-year period. The

Standard gives values for a range of thermal conductivities of insulating material from 0.02 W/mK to 0.07 W/mK for ductwork and from 0.025 W/mK to 0.045 W/mK for pipes. The following notes and Table 11.2 are based on one, average, value. Guidance should be sought from BS 5422 if materials with other values are to be used.

The economic thickness for insulation on ductwork carrying warm air depends on the temperature difference between the air inside and outside the duct and is as follows, based on material with a thermal conductivity of 0.04 W/mK:

Temperature difference (K)	Thickness of insulation (mm)
10	38
25	50
50	63

The economic thickness for pipe insulation for heating and hot water systems is shown in Table 11.2 based on insulation with a thermal conductivity of 0.035 W/mK. As can be seen different thicknesses apply depending which system is being insulated and whether the pipe is running through a heated or unheated area.

Table 11.2 also shows the same insulating material thicknesses for cold water pipes necessary to give protection against freezing. The Standard points out that if the ambient temperature remains low for a long time and there is no movement of water along the pipe, the use of insulation alone will not afford a complete protection against internal freezing. For the smaller pipes it is not practicable to install thermal insulation of sufficient thickness to avoid entirely the possibility of ice formation overnight in subzero temperatures. The problem is more acute the smaller the pipe, due to the lesser heat capacity, and, as can be seen from the table, the thickness required to afford reasonable protection to a 15 mm pipe in an outside location is a totally impracticable

Table 11.2 Minimum thickness of pipe insulation (in millimetres) for insulation with a thermal conductivity of 0.035 W/mK

Pipe size	Central heating		Hot water		Cold water	
	Heated areas	Unheated areas	Heated areas	Unheated areas	Indoors	Outdoors
15	21	32	14	17	62	279
22	30	34	16	20	20	47

(Source: Extracted from BS 5422: 1990)

279 mm. Changing the pipe size to 22 mm reduces this to a realistic 47 mm and is a recommended practice, unless some form of suitable pipe heating is employed.

11.10.4 Alternative approach

The requirements of the Regulation can be met by adopting the recommendations contained in:

- BS 5449: 1990 *Specification for forced circulation hot water central heating systems for domestic premises,* or
- BS 5864: 1989 *Specification for installation in domestic premises of gas-fired ducted air-heaters of rated output not exceeding 60 kW.*

The recommendations followed must, however, contain zoning, timing and anti-cycling features that are similar to those described above.

Chapter 12

BATHROOMS, TOILETS AND ABOVE GROUND DRAINAGE

| 12.1   The Building Regulations applicable |

F1   Means of ventilation
G1   Sanitary conveniences and washing facilities
G2   Bathrooms
H1   Sanitary pipework and drainage D12.1

F1 requires that there be adequate ventilation for the people in the building, specifically in this case in the Bathroom and Toilet; G1 states that there must be adequate sanitary conveniences (by this is meant water and chemical closets) and washbasins, with hot and cold water, in rooms separated from places where food is prepared; G2 requires that a bathroom with a fixed bath or shower, supplied with hot and cold water, should be provided; and H1 is concerned that any system carrying foul water from appliances should be adequate. D12.2

In this chapter the more commonly used name of toilet is employed for what is described in the Regulations as Sanitary Accommodation and water closet or WC is used for the fitting, although the Regulations do allow the use of chemical closets in buildings (subject to the same rules) where there is no suitable water supply or means of disposal of foul water.

| 12.2   Sanitary provisions generally |

Approved Document G1 states that the requirements of Regulation G1 will be met by a sufficient number of sanitary conveniences of the appropriate type for the sex and age of the people using the building. In a house, flat or maisonette the sufficient number is considered to be at least one WC and one washbasin. A house in multiple occupation, that is, one in which the occupants do not form part of the same household, should also have at least one WC and one washbasin with the provision that they should be accessible to all occupants. D12.3

The washbasin should be in the same room as the room containing the WC, or in a space adjacent to it, sited, designed and installed so as not to be prejudicial to health.

Similarly, any house, flat or maisonette should have at least one bathroom containing a fixed bath or shower and a house in multiple occupation should have the same minimum provision accessible to all the occupants.

The washbasin and the bath or shower must have a supply of hot water which may be from a central source or from a unit water heater as well as a piped supply of cold water.

The design of a WC and a washbasin should provide a surface which is smooth, non-absorbent and easily cleaned The flushing apparatus of a WC should be capable of cleansing the receptacle effectively. and no part of the receptacle should be connected to any pipe other than a flush pipe or a soil pipe branch.

12.2.1 Materials for pipework

Approved Document H1 lists the following British Standards specifying materials which are suitable for use in sanitary pipework:

Use	Material	British Standard
Pipes	cast iron	BS 416 and BS 6087
	copper	BS 864 and BS 2871
	galvanized steel	BS 3868
	uPVC	BS 4514
	polypropylene	BS 5254
	plastics	BS 5255
	ABS	
	MUPVC	
	polyethylene	
	polypropylene	
Traps	copper	BS 1184
	plastics	BS 3943

If different metals are used they should be separated by non-metallic material to prevent electrolytic action and all pipes should be firmly supported without restricting thermal movement.

12.2.2 Airtightness

All pipes, fittings and joints should be capable of withstanding an air or smoke test with a positive pressure of at least 38 mm water gauge for at least 3 minutes. During this time, every trap should maintain a water seal of at least 25 mm. Smoke testing is not recommended for uPVC pipework.

12.3   Bathrooms

In addition to an opening window, of no specific size, a bathroom must be fitted with background ventilation of 4000 mm^2 and an extract fan with a performance of 15 l/s. An alternative to the extract fan would be passive stack ventilation by a duct from the ceiling level to a roof terminal. If the bathroom is an internal room and cannot have an opening window, the extract fan should be operated through the light switch and be fitted with a 15 minute overrun. There should also be an air inlet such as a 10 mm gap below the door.

The bath or shower on an upper floor should discharge through a grating and trap to a branch waste pipe connected to a soil pipe. If the bathroom is on the ground floor, the bath or shower can be connected directly to a soil drain or discharge into a gully. The traps, waste pipes, soil pipes and gullies should all be provided in accordance with Approved Document H1 and Section 12.6 of this chapter.

12.4   Toilets

A room containing a WC should be provided with ventilation comprising either an opening or openings with a total area equal to one-twentieth of the floor area of which some part must be at least 1.75 m above the floor, or an extract fan with a performance of 6 l/s. In addition the room should be provided with background ventilation of at least 4000 mm^2. If it is an internal room with no openable window, an extract fan must be provided as above, controlled through the light switch and with an overrun arrangement that ensures that the fan continues for at least 15 minutes after the light has been switched off. An alternative is to provide passive stack ventilation as for bathrooms. Whether a fan or PSV is installed, an air inlet must be provided such as a 10 mm gap under the door. The room should also be separated, by a door, from a kitchen or space where food is prepared and anywhere where washing up is done.

There should also be a washbasin in the room containing the WC or else in the adjoining room (see Fig. 12.1).

It should be noted that, whilst the plan arrangements shown in Fig 12.1 satisfy the requirements of the Building Regulations, it may be found that the local authority would not approve it under the provisions of the Housing Act, Section 604. **D12.4**

The WC should discharge through a trap and branch pipe to a soil pipe or, if it is on the ground floor, directly to a soil drain. The washbasin should discharge through a grating and trap to a branch pipe connected to a soil pipe or, if it is on the ground floor, to a gully or directly to a soil drain. All pipework to be in accordance with Approved Document H1 and Section 12.6 of this chapter.

Any room including a Kitchen

Any room including a Kitchen

Any room except a Kitchen

Fig. 12.1 Plans of WC compartments

12.5   Macerators and pumps

Sanitary fittings may be connected to a macerator and pump where there is inadequate fall for a gravity system. The discharge from the pump may be taken through a small bore waste pipe to a soil pipe.

The macerator, pump and small bore system must be the subject of a current European Technical Approval issued by a member body of the European Organization for Technical Approvals, such as the British Board of Agrément and the conditions of use are in agreement with the terms of that Approval document.

If it is the WC which is connected to the macerator and pump, there must be another WC, connected to a gravity system, accessible in the building.

12.6   Sanitary pipework

Pipe sizes given in Approved Document H1 and in this section are nominal sizes in round figures approximating to the manufactured sizes. Equivalent pipe sizes for individual pipe standards can be found from the manufacturer or in BS 5572.

The capacity of the pipework system should be large enough to carry the expected flow at any point. This capacity depends on the size and gradient of the pipes and the flow depends on the type, number and grouping of the appliances. The minimum pipe sizes in common use are capable of carrying the flow from quite large numbers of appliances.

As appliances are seldom used simultaneously, the flow rate to be accommodated is not the total of their respective discharges. Flow rates resulting from the typical household group of one WC, one bath, one or two washbasins and one sink, as used for design purposes in BS 8301 are as follows:

Number of dwellings (houses, flats, etc.)	Flow rate (l/s)
1	2.5
5	3.5
10	4.1
15	4.6
20	5.1
25	5.4
30	5.8

12.6.1 Traps

A trap or water seal should be fitted at all points of discharge into the drainage system to prevent foul air from entering the building. Under working conditions and under test a trap should retain a minimum seal of 25 mm. Minimum trap sizes and depths of seal are as follows:

Appliance	Diameter of trap (mm)	Depth of seal with discharge to soil pipe (mm)	Depth of seal with discharge to gully (mm)
Washbasin	32	75	75
Bidet	32	75	75
Sink	40	75	38
Bath	40	75	38
Shower	40	75	38
Food waste disposal unit	40	75	75
WC pan	75 (siphonic only)	50	not applicable

All traps should be accessible for the purpose of clearing blockages. If the trap forms part of a fitting, the whole appliance must be removable, otherwise the trap should be fitted directly after the appliance and should be either removable or fitted with a cleaning eye.

12.6.2 Branch pipes

Branch pipes from appliances on upper floors should discharge into another branch pipe or a soil stack. If the appliance is on the ground floor the branch pipe may be connected to a soil stack, a stub stack, or directly to a drain. If the pipe carries only waste water it may, alternatively, discharge into a gully, terminating between the grating or sealing plate and the surface of the water seal in the trap.

The branch connections to a stack pipe should be offset to avoid causing crossflow into other branch pipes and, in single dwellings up to three storeys, they should not discharge into a stack lower than 450 mm above the invert of the tail of the bend at the foot of the soil stack as shown in Fig. 12.2.

With buildings up to five storeys, the height of a branch above the tail of the bend should be not less than 750 mm, over five storeys and up to twenty storeys the ground floor appliances should discharge into their own stack and in buildings over twenty storeys, both the ground and first floor appliances should discharge into their own soil stack.

A branch pipe from a WC should only discharge directly into a drain if the drop from the crown of the trap to the invert of the drain is less than 1.5 m.

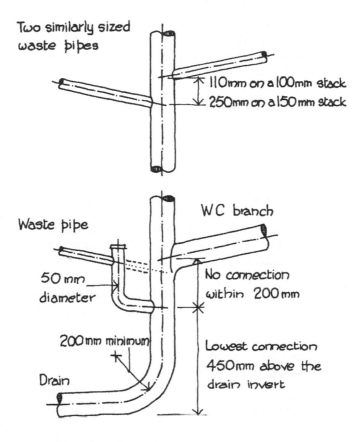

Two similarly sized
waste pipes

110mm on a 100mm stack
250mm on a 150mm stack

Waste pipe

WC branch

50 mm
diameter

No connection
within 200mm

200mm minimum

Drain

Lowest connection
450mm above the
drain invert

Fig. 12.2 Connection of branch pipes to soil pipe

The size of a branch pipe serving a single fitting should be not less than the diameter of the trap to which it is connected. If it serves more than one appliance, the minimum size for up to eight WCs should be 100 mm with a fall of between 9 and 90 mm/m and for up to four washbasins it should be 50 mm with a fall of between 18 and 45 mm/m. The length of the branch serving the eight WCs should not exceed 15 m and that to the four washbasins should not be greater than 4 m with no bends.

Bends in branch pipes should be avoided as far as possible. Where bends have to be fitted, they should be of as large a radius as practicable. In pipework of 65 mm diameter and under the radius of a bend should be not less than 75 mm measured to the centre line of the pipe.

Junctions should be formed with a sweep of 25 mm radius if the pipe is under 75 mm diameter and 50 mm radius if the pipe is 75 mm or more, or, in either case, connecting at 45°.

12.6.3 Ventilation of branch pipes

Branch pipes can be unvented up to the lengths shown in Fig. 12.3, beyond these limits a branch ventilating pipe must be fitted to prevent pressures developing which could cause the water seal in the traps to be lost.

The ventilating pipe can be taken to external air or to a soil stack to form a modified single stack system or to a ventilating stack to form a ventilated system. The latter is only likely to be the preferred arrangement where there are a large number of ventilated branches or the ventilating pipe runs to the soil stack are very long. It must be connected to the branch pipe within 300 mm of the trap and, if it is run into the soil stack, connected above the 'spill-over' level of the highest appliance served (see Fig. 12.4). Branch ventilating pipes run to the external air should terminate in the same manner as the ventilating part of a soil stack, i.e. at least 900 mm above any opening into the building within 3 m of the pipe.

The size of the ventilating pipe in a branch serving one appliance should be not less than 25 mm and, where the branch is longer than 15 m or has more than five bends, the ventilating pipe should be at least 32 mm.

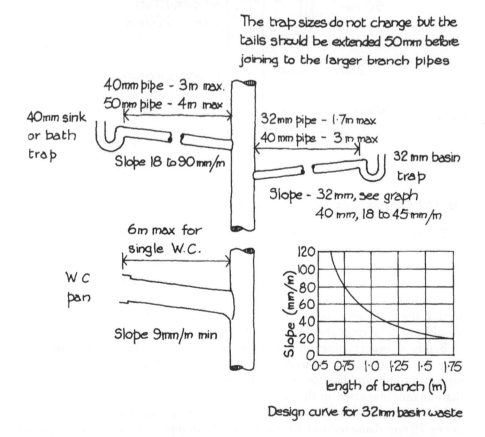

Fig. 12.3 Lengths and gradients of branch pipes

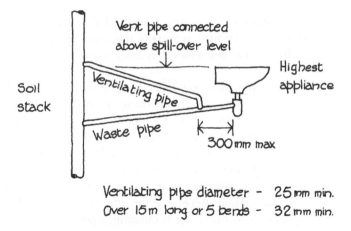

Soil stack

Vent pipe connected above spill-over level

Ventilating pipe

Waste pipe

Highest appliance

300 mm max

Ventilating pipe diameter - 25 mm min.
Over 15 m long or 5 bends - 32 mm min.

Fig. 12.4 Waste branch ventilating pipe

12.6.4 Soil stacks

In Approved Document H1 these are referred to as discharge stacks. There is no requirement given in the Approved Document as to whether to fit a soil stack inside or outside a building less than three storeys in height. Over three storeys high the soil stack should be located inside the building.

The minimum diameter for a soil stack serving appliances other than a WC is 50 mm if the flow is under 1.2 l/s and 65 mm if the flow rate is between 1.2 and 2.1 l/s. Where there is one siphonic WC only plus any other appliances, and the flow rate does not exceed 3.4 l/s, the stack size can be 75 mm. Stacks of 90 and 100 mm diameter are suitable for all types of appliance, subject to maximum flow rates of 5.3 l/s and 7.2 l/s respectively.

There should be no offsets in the 'wet' portion of a soil stack, if they can be avoided. Where it is impossible to design the system without an offset, there should be no branch connections within 750 mm of the offset. If the building is more than three storeys in height, a ventilating stack may be needed with connections above and below the offset.

All soil stacks should, of course, discharge to a drain and the radius of the bend at the foot of the stack should be not less than 200 mm to the centre line of the pipe – larger if possible.

12.6.5 Ventilation of soil stacks

Soil stacks should be ventilated for the same reason as long branch pipes, to prevent the build-up of pressures which may force the water out of the seals in the traps. Furthermore, soil stacks connected to drains which are liable to surcharging or backing up, or are connected near an intercepting trap, require ventilating pipes of not less than 50 mm diameter connected above the likely flood level of the drain.

That part of the soil stack which serves as a ventilating pipe only – the dry part above the highest connection – is usually the same diameter as the wet part of the soil pipe, but in one and two storey houses it may be reduced to not less than 75 mm.

Soil stack ventilating pipes may terminate outside or inside the building. Those terminating in the outside air should do so at least 900 mm above any openings into the building within 3 m of the pipe and should be finished with a cage or terminal which does not restrict the flow of air. Those terminating inside the building should be fitted with an air admittance valve which has a current British Board of Agrément Certificate. The use of such air admittance valves should be carefully monitored to ensure that they do not adversely affect the amount of ventilation necessary for the below ground drainage system as normally provided by open top ventilating pipes.

12.6.6 Stub stacks

Unventilated stub stacks may be used for above ground drainage within certain limits. The stub stack should connect into either a ventilated soil stack or directly into a drain. Branch pipes taking waste water only should be connected to the stub stack within 2 m of the connection to the ventilated stack or drain. Branches which serve a WC should connect to the stub stack within 1.5 m, measured from the invert of the drain or connection to the ventilated stack to the crown of the WC trap.

The length of branch drain from a stub stack should be limited to 6 m where a single appliance is connected and 12 m where a group of appliances is served.

12.6.7 Access

It must be possible to gain access to all parts of the above ground system for the purpose of clearing any blockages that might occur.

Rodding points should be fitted in any length of discharge pipe that cannot be reached by the removal of a trap and also where necessary to give access to any length of pipe that cannot be reached from any other part of the system.

In addition, all pipes should be reasonably accessible to allow any essential repairs to be carried out.

12.6.8 Pipes penetrating compartment floors or walls

Any soil pipe, waste pipe or any other pipe which penetrates a floor or wall which separates dwellings from each other or from an escape route creates a fire hazard and in many situations would also nullify the sound insulation provisions made. In these circumstances, the pipes must be enclosed in a duct or a sleeve pipe and fire stopping constructed as described in Section 5.5 of

Chapter 5 in respect of floor penetration and Section 7.3.2 of Chapter 7 with respect to wall penetration.

12.6.9 Draught sealing of services

Air leakage at the point where pipes and other services penetrate or project into hollow constructions or voids has been identified as a source of heat loss. To minimize this Approved Document L requires that all such services are sealed at the point of penetration. This means that any holes in ceilings or walls through which pipes (and electrical cables) pass need to filled after the service has been installed with a suitable sealant such as a mastic.

12.7 Alternative approaches

As an alternative to the detailed recommendations given in the Approved Documents and set out above, the requirements of the Regulations can be met by adopting the relevant recommendations of the following British Standards:

- BS 6465 *Sanitary installations,* Part 1: *Code of practice for scale of provision, selection and installation of sanitary appliances* (clauses 2, 3, 6, 7 and 8 are relevant).
- BS 5572: *Code of practice for sanitary pipework* (clauses 3, 4 and 7 to 12 are relevant).

Chapter 13

BELOW GROUND DRAINAGE, RAINWATER DISPOSAL AND SOLID WASTE STORAGE

| 13.1   The Building Regulations applicable |

The Building Regulations relating to the disposal of the waste products of a residence are:

B1   Means of escape
E1   Airborne sound
H1   Sanitary pipework and drainage
H2   Cesspools and tanks
H3   Rainwater drainage
H4   Solid waste storage

The relevant part of Regulation B1 is concerned with the positioning of refuse chutes in relation to escape routes from buildings; E1 sets standards of sound insulation to be achieved between a habitable room and a refuse chute; H1 requires that any system which carries foul water should be adequate; H2 states that cesspools and the like should be of adequate capacity and correctly constructed and sited; H3 requires that any system which carries rainwater must be adequate; and H4 calls for a similar adequacy in the storage of solid waste.

| 13.2   Drain design and layout |

The design of the drainage system will depend in the first instance on whether it will discharge into a public sewer or an on-site disposal method and, if there is a public sewer, whether it is for soil drainage only or a combined sewer taking both soil drainage and rainwater. The latter is less common at the present time but if it is available, the pipes for the drainage must be sized to take account of the possibility of the peak flow rates of both the soil drainage and the rainwater drainage occurring at the same time. If the discharge is to a cesspool or septic tank a combined system cannot be used.

The normal, and preferable, arrangement is to design the system to fall all the way to the public sewer but, if such a gravity system is not possible or is impracticable, sewage lifting equipment would have to be employed and should be installed in accordance with the recommendations contained in BS 8301: *Code of practice for building drainage.*

The connection between drain runs and the sewer should be made obliquely and in the direction of flow. The simpler the layout of the system the better it will work, changes of direction and gradient should be as few as possible and as easy as practicable. Access points should be provided only where they are necessary to allow a blockage of the drain to be cleared.

The system needs to be ventilated by a flow of air and a ventilating pipe should be provided:

1 at or near the head of each main drain;
2 at the head of any branch longer than 6 m which serves a single appliance;
3 at the head of any branch longer than 12 m serving a group of appliances;
4 on any drain fitted with an intercepting trap, particularly in a sealed system.

Pipes should be laid to even gradients and in straight lines as far as practicable. Where essential, bends of as large a radius as practicable may be used in the drain run, provided that any possible blockages can still be cleared and they should be restricted to a position close to an access point or at the foot of a soil pipe.

13.2.1 Materials for pipes and jointing

Any of the materials which comply with the British Standards set out below are acceptable:

Material	British Standard
Rigid pipes:	
asbestos	BS 3656
vitrified clay	BS 65, BSEN 295
concrete	BS 5911
grey iron	BS 437
Flexible pipes:	
uPVC	BS 4660 and BS 5481

The joints should be formed in material appropriate to that of the pipe and to minimize the effect of differential settlement, all joints should be flexible Nothing should project into the drain whereby a blockage could be created.

13.2.2 Watertightness

When the drain has been laid, haunched or surrounded and backfilled up to 300 mm it should be tested by a water test or an air test.

The finished drain should be capable of withstanding a water test pressure produced by 1.5 m head of water above the invert of the pipe at the head of the drain. If the test is applied by a stand pipe of the same diameter as the drain, it should be filled, left for 2 hours and then topped up. The leakage over the next 30 minutes should then be measured and, in a 100 mm drain, it should not exceed 0.05 litres for each metre run, which is equivalent to a drop in water level of 6.4 mm per metre, or, in a 150 mm drain, it should not exceed 0.08 litres which is equivalent to a 4.5 mm/m drop. With a long run of drain the amount of the fall added to the test head can induce a pressure capable of causing damage. To prevent this, the test may need to be applied to sections of the drain so that the maximum total head does not exceed 4 m.

An air test is applied using a manometer which is mainly a U-tube containing water graduated in millimetres. As air is pumped into the drain the water level in the U-tube rises to indicate the pressure and pumping continues until the water level reaches either the 50 mm or the 100 mm mark. This is left for 5 minutes, after which time the water level should not have dropped more than 25 mm below the 100 mm head or 12 mm below the 50 mm head.

13.3   Pipe sizes and gradients

The drains must be capable of carrying the peak flow anticipated. This peak flow depends on the number and type of appliance discharging into the drain but in domestic work it is invariably a group of fittings, typically a WC, a bath, one or two washbasins and a sink. The flow rate is not the total discharge from such a group because all the appliances are seldom used simultaneously and for design purposes the flow rates for such a group of appliances can be taken as:

Number of dwellings	Flow rate (l/s)
1	2.5
5	3.5
10	4.1
15	4.6
20	5.1
25	5.4
30	5.8

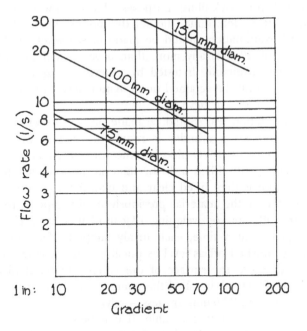

Values shown are for pipes
running three quarters full

Fig. 13.1 Discharge capacities for soil drains

The capacity of the drain depends on both the diameter of the pipes used and the gradient at which they are laid. Figure 13.1 shows a graph of the discharge capacities of foul drains running, as shown, at 0.75 proportional depth. From this it can be seen that the minimum gradient for a 75 mm pipe is 1 in 80 giving a capacity of 2.8 l/s, for a 100 mm pipe it is also 1 in 80 (subject to there being at least one WC discharge) and would give a capacity of 6.3 l/s and for a 150 mm pipe the minimum gradient is 1 in 150 (provided that there are at least five WCs connected to it) and its capacity would be 15.0 l/s. The minimum WC requirement is because a drain pipe needs to have an adequate flow through it to keep it clean, particularly at the flatter gradients where the effluent has a low velocity.

13.4 Bedding and backfilling

The choice of bedding and backfilling materials depends on the depth at which the pipes are laid, their size and strength and the nature of the ground excavated. The last of these is involved where it is possible to re-use the excavated material for backfilling purposes. The following on-site test is described in BS 8301 to determine whether the 'as-dug' material is suitable.

Firstly, a visual examination should be made to see whether there are any particles in excess of 20 mm, any more than a very small proportion would render the material unsuitable. If tested by a sieve, a 2 kg sample should be taken and nothing should be retained on a 38 mm sieve with no more than 5 per cent by mass on a 19 mm sieve.

Secondly, a compaction test should be made on a representative sample of about 11 kg. To do this, a cylinder is required, 150 mm in diameter and 250 mm long (a length of uPVC pipe is quite suitable), placed on end on a smooth, flat surface, gently filled with the material poured in until the top is level across. The pipe is then lifted clear and placed on another smooth, flat surface. One-quarter of the material previously in the pipe is returned to it and tamped with a 40 mm rod weighing 1 kg until it is firm. This is followed by the rest, a quarter at a time, each firmly tamped. The top of the final quarter should be finished off as level as possible. The distance from the top end of the pipe down to the surface of the tamped material divided by the length of the pipe (250 mm) is called the compaction fraction.

If the compaction is 37.5 mm or less – giving a maximum compaction fraction of 0.15 – the material is suitable. If it is between 37.5 and 75 mm – giving a maximum compaction fraction of 0.30 – the material may be suitable in normal ground conditions with care but it would be unsuitable if there was a possibility of the trench getting waterlogged. Material with a compaction fraction over 0.30 is not suitable for backfilling.

The appropriate granular material for backfilling and bedding is determined by the pipe size as follows:

Pipe size (mm)	Bedding material
110	10 mm nominal single sized aggregate
160	10 mm or 14 mm nominal single sized aggregate or 14 mm to 5 mm nominal graded aggregate
225 and over	10 mm, 14 mm or 20 mm nominal single sized aggregate or 14 mm to 5 mm or 20 mm to 5 mm nominal graded aggregate

Suitable arrangements for bedding and backfilling of both rigid and flexible pipes are shown in Fig. 13.2. In detail A, the barrel of the pipe should be supported by the bottom of the trench which must be very accurately trimmed to the falls by hand and with suitable pockets taken out to receive the collars. This is only suitable where the trench is in ground that would pass a compaction fraction test.

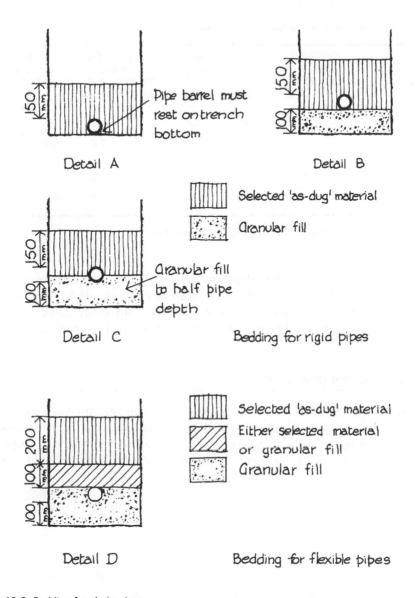

Fig. 13.2 Bedding for drain pipes

Table 13.1 Limits of cover for rigid pipes (m)

Pipe bore	Bedding detail (Fig. 13.2)	Fields and gardens		Light traffic roads		Heavy traffic roads	
		Min.	Max.	Min.	Max.	Min.	Max.
100	A or B	0.4	4.2	0.7	4.1	0.7	3.7
	C	0.3	7.4	0.4	7.4	0.4	7.2
150	A or B	0.6	2.7	1.1	2.5	—	—
	C	0.6	5.0	0.6	5.0	0.6	4.6

(*Source:* Based on Table 8 of Approved Document H1)

The minimum and maximum depth of cover for rigid pipes is shown in Table 13.1. For flexible pipes the maximum depth of cover should be 10 m and the minimum 0.9 m under any road or 0.6 m under any other surface. In both cases the minimum can be reduced if special protection is provided.

13.5 Special protection

There are situations where the standards set out in the Approved Document cannot reasonably be achieved or where a particular set of circumstances requires extra precautions to be taken. To allow for this the Approved Document describes a number of special protection measures that can be taken.

13.5.1 Pipes under buildings

Drains may be run under buildings in the normal way, provided that there is at least 100 mm of granular fill around the pipes unless the crown of the pipe is within 300 mm of the underside of the ground floor slab, in which case concrete encasement, integral with the slab, should be used.

If there is the possibility of excessive settlement occurring additional flexible joints may be required or, alternatively, the drain could be suspended.

13.5.2 Pipes through walls

To accommodate any possible movement of the wall *one* of two options should be adopted:

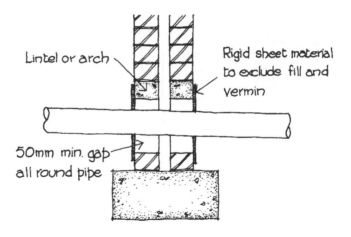

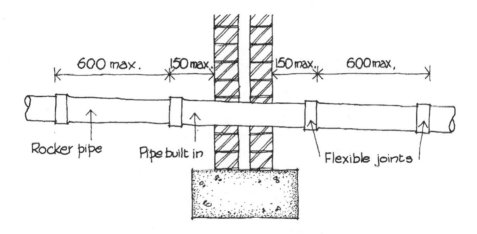

Fig. 13.3 Drain pipes passing through walls

1 An opening is formed in the wall leaving at least 50 mm clearance all round the pipe, with masking sheets fitted round the pipe against both wall faces to close off the opening against possible ingress by vermin.
2 A pipe is built into the wall of such a length that its ends are just clear of the face and to these are connected flexible jointed rocker pipes, each no more than 600 mm long, which in turn, are connected to the drain runs (see Fig. 13.3).

13.5.3 Pipes near walls

A drain trench should not be excavated below the level of the foundations of any nearby building unless either of the following conditions exists:

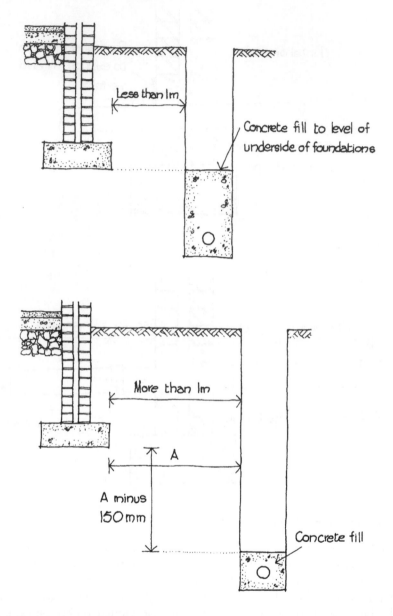

Fig. 13.4 Drains near to building foundations

1 Where it is within 1 m of the foundation the trench is filled with concrete up to the level of the underside of the foundations.
2 Where it is more than 1 m from the foundation, the trench is filled with concrete to a point which is lower than the underside of the foundations by a height equal to the distance from the building minus 150 mm (see Fig. 13.4).

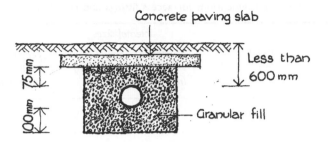

Fig. 13.5 Minimum protection for flexible pipes

13.5.4 Reduced amount of cover

Where it is not possible to provide the amount of cover to rigid pipes shown in Section 13.4, the drain should be encased in not less than 100 mm of concrete with movement joints formed with compressible board at each socket.

Flexible pipes with less than 600 mm of cover should be laid as normal, with at least 100 mm of granular fill below the pipe and 75 mm above. This is then protected by concrete slabs laid to bridge the filling (see Fig. 13.5).

13.5.5 Surcharging of drains

Where a drain is liable to surcharging or flooding, measures should be taken to protect the building. The precise action to be taken depends on the circumstances and guidance on protective measures is to be found in BS 8301. Where any type of anti-flood device is installed, it may be necessary to provide additional ventilation of the drain to avoid the trap seal being lost.

13.5.6 Rodent control

In certain areas the infestation of the drains and private sewers can be a problem. Approved Document H1 suggests that the local authority can provide the best information on the protective measures found to be the most effective in their particular area.

13.6 Access

To make it possible to clear any blockages that might occur in the drains, sufficient and suitable access points are to be provided, of a type sited and spaced appropriate to the layout of the system, its depth and the size of the drain runs.

The Approved Document covers the normal method of clearance by rodding and not the use of any mechanical means.

Table 13.2 Minimum dimensions for access fittings and chambers

Type	Depth (m)	Internal sizes		Cover sizes	
		Length × width (mm × mm)	Circular (mm)	Length × width (mm × mm)	Circular (m)
Rodding eye		As drain but minimum 100			
Access fitting:	0.6 or less				
small		150 × 100	150	150 × 100	150
large		225 × 100	—	225 × 100	—
Inspection chamber	0.6 or less	—	190*	—	190*
	1.0 or less	450 × 450	450	450 × 450	450†
Manhole	1.5 or less	1200 × 750	1050	600 × 600	600
	over 1.5	1200 × 750	1200	600 × 600	600
	over 2.7	1200 × 840	1200	600 × 600	600
Shaft	over 2.7	900 × 840	900	600 × 600	600

(*Source:* Based on Table 9 of Approved Document H1)

* Drains up to 150 m.
† Clayware or plastics may be reduced to 430 mm to support cover frame.

Four types of access points are covered, rodding eyes, access fittings, inspection chambers and manholes, which are defined as:

- Rodding eyes: capped extensions of the pipes.
- Access fittings: small chambers on (or an extension of) the pipes, but not with an open channel.
- Inspection chambers: chambers with working space at drain level.
- Manholes: large chambers with working space at drain level.

The depths at which each type should be used and the dimensions recommended are shown in Table 13.2.

13.6.1 Siting and spacing of access points

Access points are required at:

- the head of the drain;

Table 13.3 Maximum spacing of access points (m)

From	To	Access fitting		Junction	Inspection Chamber	Manhole
		Small	Large			
Start of drain		12	12	—	22	45
Rodding eye		22	22	22	45	45
Access fitting:						
150 diam		—	—	12	22	22
150 × 100		—	—	12	22	22
225 × 100		—	—	22	45	45
Inspection chamber		22	45	22	45	45
Manhole		22	45	45	45	90

(*Source:* Based on Table 10 of Approved Document H1)

- a bend;
- a change of gradient;
- a change of pipe size;
- a junction (unless each run can be cleared from another access point).

In addition there should be access points in the length of any long runs of drain as shown in Table 13.3.

13.6.2 Construction of access points

All access points must be capable of containing the contents of the drain both under test and under working conditions. In addition, they should be able to resist ground water and rainwater finding their way in.

Materials conforming to the following British Standards are suitable for use as inspection chambers or manholes.

Rodding eyes and access fittings should be of the same material as the pipes.

Inspection chambers and manholes with open channels should have the branches fitted so that they discharge at or above the level of the top edge of the channel and wherever the angle of the branch is more than 45° a three-quarter channel section should be used.

Benching of the main channel and the branches should be at a slope of 1 in 12 and carried up to at least the top of the outgoing pipe. It should be finished at the channel with a rounded edge of at least 25 mm radius.

Material	British Standard
Clay:	
bricks and blocks	BS 3921
vitrified	BS 65
Concrete:	
precast	BS 5911
in situ	BS 8110
Plastics	BS 7158

A non-ventilating cover must be fitted to all inspection chambers and manholes. It must be of a durable material, such as cast iron, cast or pressed steel, precast concrete or uPVC and of a strength suitable for its position. Covers inside buildings should be mechanically fixed and airtight unless the drain itself has a watertight sealed access cover.

Manholes deeper than 1 m should be fitted with step irons or a fixed ladder.

13.7   Cesspools and tanks

Specialist knowledge is advised in Approved Document H2 in the design and installation of small sewage treatment works and guidance can be found in BS 6297: *Code of Practice for design and installation of small sewage treatment works and cesspools.*

The capacity of a cesspool should be sufficient to enable it to store all the soil drainage of the building until it can be emptied. The Approved Document states that this should be at least 18 000 litres or 18 m^3 below the level of the inlet.

The capacity of septic tanks and settlement tanks needs to be sufficient to allow the breakdown of solid matter in the soil drainage and should be at least 2700 litres or 2.7 m^3 below the level of the inlet.

If, as is usual, a cesspool, septic tank or settlement tank is intended to be desludged using a tanker, it should be sited within 30 m of a suitable vehicle access. It should be arranged so that the emptying, desludging and cleaning can be carried out without any hazard to the occupants of the building and without any of the contents having to be taken through either a dwelling or a place of work. Access may, however, be through an open covered space.

13.7.1   Design and construction

Cesspools should be designed so that they are:

1 covered with heavy concrete slabs and ventilated;
2 provided with an access for emptying;
3 provided with access for inspection at the inlet;
4 have no openings other than for the inlet, access for emptying, ventilation and inspection.

Septic tanks and settlement tanks should be designed so that they are:

1 covered with heavy concrete slabs and ventilated or fenced in;
2 provided with access for emptying, desludging and cleaning;
3 provided with access for inspection at both the inlet and the outlet.

Septic tanks should have two chambers operating in series and the inlet and the outlet should be designed to prevent disturbance of the surface scum or settled sludge. Where the width of the chamber does not exceed 1200 mm the inlet should be via a dip pipe. If the incoming drain is laid to a steep gradient, the velocity of the effluent can cause turbulence. To reduce this, the last 12 m should be laid at a gradient of no more than 1 in 50.

The construction of cesspools and tanks can be in engineering bricks laid in 1:3 cement and sand mortar to a thickness of at least 220 mm; concrete at least 150 mm thick of C/25/P mix as specified in BS 5328; glass reinforced concrete or factory made in glass reinforced plastics, polythene or steel provided that they carry a British Board of Agrément Certificate and are installed in strict accordance with both the certificate and the manufacturer's instructions. The Approved Document points out that particular care is needed to ensure the stability of factory made cesspools and tanks.

The access to a cesspool or tank should have no dimension less than 600 mm where access is required and the cover should of a durable material consistent with the corrosive nature of the contents of the tank. All covers should be lockable.

13.8 Rainwater disposal

Approved Document H3 gives guidance on ways to meet the requirements for the drainage of rainwater from roofs of over $6\,m^2$ area and it assumes an intensity of fall of 75 mm/h.

13.8.1 Gutters

The material of a gutter should be of adequate strength and durability, all joints should remain watertight under working conditions and the gutter should be firmly supported without restricting any thermal movement. Any differing metals should be separated by non-metallic material to prevent electrolytic action.

The size of a gutter must be adequate to carry the expected flow at any point in the system. This is calculated in relation to the 'effective area' which relates to the actual area of roof to be drained and whether it is flat or pitched plus, if it is pitched, the angle of pitch. For a flat roof the effective area is the same as the actual area. For a pitched roof, the effective area is calculated as shown below:

30° pitch	plan area × 1.15
45° pitch	plan area × 1.40
60° pitch	plan area × 2.00
70° pitch and over	elevational area × 0.5

Using these effective areas the gutter size can be found by reference to Table 13.4 below. The gutter referred to in this Table is half round in section, of a length from stop end to outlet less than 50 times the water depth, laid level with an outlet at one end. A gutter with an outlet at each end can have a length of 100 times the water depth. Where the outlet is not at the end, the gutter size should be selected to suit the larger of the roof areas draining into it.

Table 13.4 Gutter and outlet sizes

Effective roof area (m²)	Gutter size (mm dia)	Outlet size (mm dia)	Flow capacity (l/s)
6.0 to 18.0	75	50	0.38
19.0 to 37.0	100	63	0.78
38.0 to 53.0	115	63	1.11
54.0 to 65.0	125	75	1.37
66.0 to 103.0	150	89	2.16

(Source: Based on Table 2 of Approved Document H3)

Outlets with rounded edges have a better flow pattern which will allow the downpipe to be reduced in size. Gutters laid to fall towards the nearest outlet or which have a capacity greater than that of a half round section can be reduced in size as recommended in BS 6367: *Code of practice for drainage of roofs and paved areas*, or by following the manufacturer's recommendations, provided that they comply with BS 6367.

Gutters should also be installed so that if there is a rainfall greater than normal, the excess water in the gutter can overflow in such a manner that it will be discharged clear of the building.

13.8.2 Rainwater pipes

As with gutters, the materials used for rainwater pipes should be of adequate strength and durability and firmly fixed without restricting thermal movement. All pipe joints inside a building should be capable of withstanding the airtightness test described in this chapter for underground drainage.

The size of the pipe should be not less than the size of the gutter outlet and where more than one gutter is connected to a pipe, it should have a cross sectional area at least as great as the combined areas of the outlets.

The discharge of a rainwater pipe may be into a drain, into a gully, into another gutter or on to a drained surface. If the drain into which the pipe discharges is part of a combined system, the connection must be through a trap.

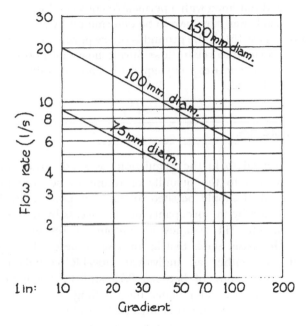

Values given are for
drains running full

Fig. 13.6 Discharge capacities of rainwater drains

13.8.3 Rainwater drainage

Generally the materials, design, layout, and installation of rainwater drains should follow the recommendation for soil drains.

The size of the drain should be sufficient to carry the anticipated flow from the roof plus any run-off from paved areas (although this run-off is not covered by the Building Regulations). The paved area run-off should be calculated on the basis of a rainfall intensity of 50 mm/h.

The minimum size of a rainwater drain is 75 mm and the minimum fall for both 75 mm and 100 mm drains is 1 in 100. Figure 13.6 shows the discharge capacities of rainwater drains running full.

Rainwater drains may be run to their own sewer, a ditch or a soakaway. The sewer may be combined and so will take soil drainage as well, but the local authority must be consulted. Any discharge to a ditch must receive the approval of the local water authority. It is quite common to use a soakaway but Approved Document H gives no guidance on their construction or size. The Approved Document refers to BS 8301 as a guide, and the Standard recommends that soakaways should not be less than about 5m from the building and may take the form of an unlined pit filled with hardcore, or seepage trenches. Larger pits may be used and left unfilled but lined with a permeable construction.

The size of the soakaway may have to be determined by a permeability test as described in BRE Digest 151. If the ground is of low permeability, suitable storage capacity must be provided to retain flows during heavy periods of rain, and the Standard recommends that a capacity equal to 12mm of rainfall over the area drained should be adopted.

13.9 Refuse storage

Refuse storage, referred to in the Regulations as solid waste storage, should have a capacity related to the quantity of refuse to be contained and the frequency of removal. It should be sited so that it is not prejudicial to health and convenient for use by the occupants and for access by refuse collectors.

Approved Document H4 assumes an output of refuse of $0.09\,\text{m}^3$ per dwelling and, on this basis, states that in low rise domestic developments each dwelling should have either an individual movable container of at least $0.12\,\text{m}^3$ or a communal one of between 0.75 and $1\,\text{m}^3$ volume, with close fitting lids. In multistorey developments, the dwellings up to the fourth floor can be treated as low rise and any over this level should have access to a refuse chute which discharges into a shared communal container. If it is impracticable to provide refuse chutes, the local authority will require to be assured of a satisfactory management arrangement for conveying refuse to the storage point.

13.9.1 Refuse chutes

Refuse chutes should be designed in accordance with BS 5960: *Code of practice for storage and on-site treatment of solid waste from buildings.* They should have a smooth, non-absorbent surface, a close fitting door at each storey where there is a dwelling and be ventilated at the top and the bottom.

Chutes should not be located within protected lobbies or stairways and be separated from the rest of the building by fire resisting and sound insulating construction.

A wall separating a refuse chute from a habitable room or kitchen should have a mass (including any finishes) of not less than $1320 \, \text{kg/m}^2$ and one separating a non-habitable room should have a mass of not less than $200 \, \text{kg/m}^2$. As a guide, a wall of mass $1320 \, \text{kg/m}^2$ can be achieved with *in situ* concrete of a density of $2200 \, \text{kg/m}^3$ and an approximate thickness of 600 mm, or a three-and-a-half brick wall with one face plastered using brickwork of a density of $1650 \, \text{kg/m}^3$.

The chute should be sited so that householders are not required to carry their refuse further than 30 m.

13.9.2 Containers

A storage chamber is not required for refuse containers but if one is provided there should be a clear space all round and between the containers of at least 150 mm and if they are communal containers the chamber should be at least 2 m high. All containers should be sited within 25 m of vehicle access and in a position where they can be collected without being taken through a building other than a garage, car port or other open covered space.

Refuse storage chambers should be separated from the other parts of the building by fire resisting construction, not located within protected lobbies or stairways and ventilated at the top and the bottom. Access to the chamber should not be adjacent to escape routes or windows of dwellings and they should be entered either directly from the open air or through a protected lobby provided with not less than $0.2 \, \text{m}^2$ of permanent ventilation.

13.10 Alternative approaches

The requirements of the Regulations in connection with soil and rainwater drainage and refuse disposal can also be met by compliance with the relevant recommendations of the following British Standards:

- BS 5572: *Code of practice for sanitary pipework.*
- BS 5906: *Code of practice for storage and on-site treatment of solid waste from buildings.*
- BS 6297: *Code of practice for design and installation of small sewage treatment works and cesspools.*
- BS 8301: *Code of practice for building drainage.*

Chapter 14

HOUSE AND BUNGALOW CONVERSIONS AND CONSERVATORIES

| 14.1   The Building Regulations applicable |

The Building Regulations which refer to the alteration of houses into flats and the conversions of lofts into habitable rooms are:

B1   Means of escape
B3   Internal fire spread (structure)
B4   External fire spread
E1   Airborne sound (walls)
E2   Airborne sound (floors)
E3   Impact sound (floors and stairs)
F2   Condensation
K1   Stairs and ramps

Much of this legislation is a relaxation of the requirements for new building, recognizing that the standards imposed for new work cannot always be reasonably applied to alteration contracts and, indeed, due to the very limited nature of the accommodation resulting in some cases, is not necessarily appropriate.

B1 contains a section specifically related to means of escape from loft conversions; B3 sets out certain conditions which, if met, can lead to the relaxation of the standard requirements for fire resistance of floors where they are in flat conversions and in loft conversions; B4 requirements with regard to rooflights apply to conversion work as well as to new; E1,E2 and E3 set out recommended constructions to improve the levels of sound insulation in existing structures; F2 requires adequate provision to be made to prevent excessive condensation in a roof; and K1 contains relaxations of the standard requirements for stairs when the flight is to a loft conversion only.

| 14.2   Flat conversions |

Generally, the standard requirements of the Building Regulations apply to all aspects of flat conversion but in the subjects of fire resistance and sound insulation, the impracticability of achieving the standards set for new work in certain circumstances is recognized and modifications are acceptable.

14.2.1 Floors – fire resistance

In Section 7 of Approved Document B3 there is an acknowledgement that if the existing timber floors are to be retained in a house to be converted into flats the standards of fire resistance may be difficult to achieve. In this case, provided that the means of escape set out in Section 2 of the same Document (detailed in Chapter 15) are provided and properly protected, the 60 minute standard of fire resistance which, normally, would be required for any floor over 5 m above the ground level can be reduced to 30 minutes provided that the building does not contain more than three storeys.

Buildings of four storeys and more, converted into flats, must comply with the full standards required by the Regulations.

14.2.2 Floors – sound insulation

It may be found that the existing floor already possesses sufficient sound insulating properties to make it acceptable. Two possibilities exist:

1 The floor is constructed in a manner very similar to those given in Chapter 5. An example of a 'very similar manner' would be if the mass of a concrete floor was within 15 per cent of the mass shown in the drawings.
2 The floor is of a type which can be shown to pass the field tests set out in BS 2750: *Measurement of sound insulation in buildings and of building elements.* This test must be carried out by an organization which is recognized and is in possession of an accreditation number issued by the National Measurement Accreditation Service (NAMAS). The local authority will require to see a test report.

Where neither of these criteria can be met the floor needs to receive treatment to improve its sound insulating properties. Three such treatments are described in the Approved Document and are shown in Figs 14.1, 14.2 and 14.3. Detail A of Fig. 14.1 shows an independent ceiling fixed below the existing floor and covered with a sound absorbing material.

The existing floorboard joints should be caulked or the whole floor overlaid with sheets of hardboard to provide the airtightness necessary for the method to work. The existing ceiling can be retained if it is lath and plaster and provides an acceptable fire resistance standard, if it is not lath and plaster it should be thickened to 30 mm by additional layers of plasterboard fixed with staggered joints.

The size of the new ceiling joists will depend on the span, bearing in mind that they can gain no support from the floor above. The appropriate size can be found from Table 6.1 of Chapter 6.

The ceiling should comprise two layers of plasterboard giving a total thickness of 30 mm, sealed around the perimeter with tape or mastic. The joints in each layer are to be staggered. Over this is to be laid a 100 mm blanket of mineral wool which has a density of at least 10 kg/m^3. The whole of this suspended ceiling must be fixed with the ceiling joists at least 25 mm clear of the existing soffit and new ceiling at least 100 mm below the existing.

The combination of the necessary depth of the new ceiling joists, their clearance below the existing ceiling and the 30 mm thickness of new ceiling results in a reduction of height in the room below of something approaching 200 mm or more. Two difficulties may arise; the room may not be high enough to be able to afford this loss of height or the lower ceiling may come below the level of the window heads. If the room is not high enough, the other two floor treatments, or the alternative suspended ceiling detail, should be considered. If the new ceiling invades the window heads, a pelmet recess can be formed by raising the construction sufficiently to clear.

An alternative treatment can be used, shown as detail A1 in Fig. 14.1, where it can be proved that none of the details A, B or C is practicable (see Figs 14.1–14.3). The construction results in less loss of ceiling height than detail A but there is also a reduction in the level of sound insulation. The difference between the two constructions is that the existing ceiling must be removed for detail A1 thus allowing the new ceiling joists to be fitted between the existing ones. They can also be supported from the existing joists by hangers of wire not more than 2 mm in diameter or metal straps not more than 25×0.05 mm. There should not be more than one fixing to every square metre of floor.

Detail B of Fig. 14.2 shows a floating floor and relies for its effectiveness on the mass of the existing structure plus the value of the resilient layer.

The existing flooring can be retained but, if it is removed, the opportunity should be taken to lay a 100 mm blanket of mineral wool on the ceiling and the new flooring should consist of boarding not less than 12 mm thick.

The existing ceiling can be retained if it is lath and plaster and can be shown to provide an acceptable level of fire resistance.

Over the existing floor is placed a resilient layer, 25 mm thick, of mineral wool with a density of between 60 and 100 kg/m^3. The lower the density, the better the insulation and where the lowest densities are used an additional support can be provided around the perimeter of the room by a timber batten fitted with a foam strip along the top and fixed to the walls.

The finished floor surface, laid over the resilient layer, can be either a timber or wood-based boarding 18 mm thick with tongued and grooved joints glued together and spot bonded to an under layer of 19 mm plasterboard, or a single or double layer of material with all joints glued, which has a total mass of at least 25 kg/m^2.

This floating layer raises the original floor level by 62 mm which means refixing skirtings and can create difficulties with doorways and the top step of a staircase.

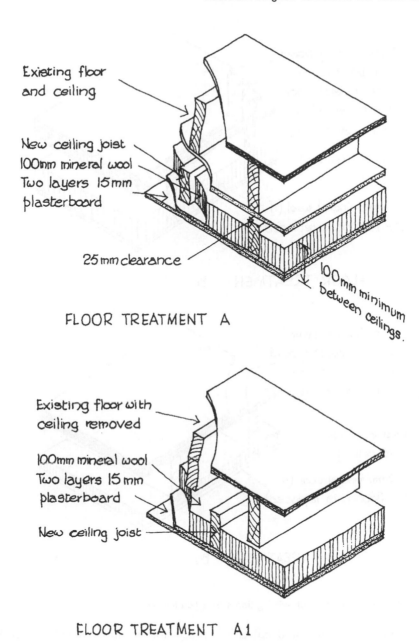

Existing floor
and ceiling

New ceiling joist
100mm mineral wool
Two layers 15mm
plasterboard

25mm clearance

100mm minimum between ceilings.

FLOOR TREATMENT A

Existing floor with
ceiling removed

100mm mineral wool
Two layers 15mm
plasterboard

New ceiling joist

FLOOR TREATMENT A1

Fig. 14.1 Sound insulation of existing floors by an independent ceiling

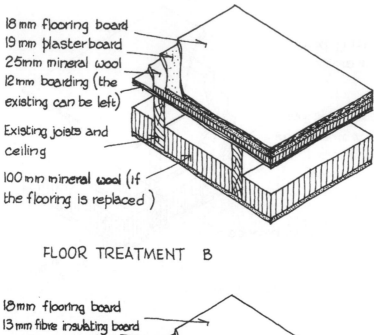

18 mm flooring board
19 mm plasterboard
25mm mineral wool
12mm boarding (the
existing can be left)

Existing joists and
ceiling

100 mm mineral wool (if
the flooring is replaced)

FLOOR TREATMENT B

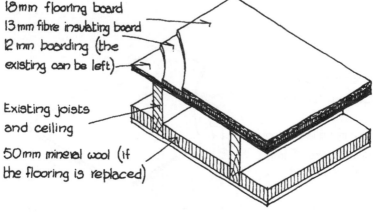

18mm flooring board
13 mm fibre insulating board
12 mm boarding (the
existing can be left)

Existing joists
and ceiling

50mm mineral wool (if
the flooring is replaced)

FLOOR TREATMENT B1

Fig. 14.2 Sound insulation of existing floors by a floating layer

If neither detail A nor detail C is a practicable solution, detail B1 can be used which only raises the floor level by half as much, or less if the existing floorboards are replaced. With this alternative the existing floorboards can be retained or can be replaced with 12 mm boarding. If they are replaced, the existing ceiling should be covered with a 50 mm blanket of mineral wool.

The existing ceiling can also be retained if it is of lath and plaster construction but, if not, it should be relined with two or three layers of plasterboard with staggered joints, giving a total thickness of 30 mm and fixed

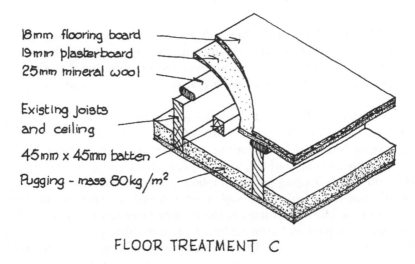

18mm flooring board
19mm plasterboard
25mm mineral wool

Existing joists
and ceiling

45mm x 45mm batten

Pugging - mass 80kg/m²

FLOOR TREATMENT C

Fig. 14.3 Sound insulation of existing floors by pugging

to timber cross-battens or suitable resilient hangers. In either case the perimeter should be sealed with tape or mastic.

The floating floor is composed of a resilient layer of fibre insulating board 13 mm thick and a floor finish of timber or wood based tongued and grooved boards 18 mm thick, glued together. There should be a gap of about 10 mm between the perimeter of the floating floor and the enclosing walls which can be filled with a resilient material.

Detail C of Fig. 14.3 does not reduce the height of the room below and can be arranged to raise the existing floor level by no more than 42 mm. It is a method that combines a floating layer and insulation or pugging within the floor.

The existing flooring must be removed and, as the new flooring will not be fixed to the existing joists, additional strutting may be needed to stiffen the timbers.

The existing ceiling can be retained if it is lath and plaster and satisfies the fire resistance requirements. If it is not lath and plaster, it should be built up to a minimum thickness of 30 mm with two or three layers of plasterboard fixed with staggered joints. The perimeter of any new ceiling should be sealed to the walls with tape or caulking.

The ceiling should then be covered with a 100 mm thick blanket of mineral wool of a density of 10 kg/m³ or pugging of a finished mass of 80 kg/m² using one of the following:

1 a 75 mm bed of traditional ash;
2 a 60 mm bed of limestone chips graded 2 to 10 mm;

3 a 60 mm bed of whin aggregate graded 2 to 10 mm;
4 a 50 mm bed of sand.

The thicknesses of pugging given will achieve the required mass but care should be taken to check and if necessary upgrade the strength of the ceiling to enable it to carry this mass. Sand is not recommended in any floors to rooms such as kitchens or bathrooms where it may possibly get wet and overload the ceiling.

Resilient strips should be laid along the tops of the floor joists consisting of mineral fibre, 25 mm thick, with a density of between 80 and 100 kg/m^3.

The floating layer which achieves the minimum increase in floor level is a single or double layer of material, with all joints glued, giving a mass of 25 kg/m^2 (two sheets of flooring grade chipboard, 18 mm thick of a density of 700 kg/m^3, would be suitable), nailed or screwed to 45 × 45 mm battens between the floor joists. Alternatively, the floating layer can be made of timber or wood-based tongued and grooved boards, 18 mm thick, glued together, spot bonded to an under layer of 19 mm plasterboard and nailed or screwed to 45 × 45 mm battens laid along the resilient strips on top of the joists.

With either type of floating layer, a gap of 10 mm should be left around the perimeter, filled with a resilient material and a gap of 3 mm left below the skirting boards.

14.2.3 Walls – fire resistance

The Building Regulations give no relaxation of the requirements for fire resistance in the walls of flats with respect to flats created by conversion. This, however, is not onerous as the standard requirements can usually be satisfactorily achieved by normal traditional construction. Houses using non-traditional systems should be investigated to check the fire resisting capabilities.

14.2.4 Walls – sound insulation

It is probable that although the wall may be satisfactory as far as fire resistance is concerned, it would not meet the sound insulating standards required. This should be checked to see whether the existing construction is generally similar to those given in Chapter 7 – for instance if the mass of the masonry of the wall is within 15 per cent of the mass of the walls shown in Fig. 7.2, in which case no further action needs to be taken. If the sound insulation does require upgrading, Approved Document E gives details of a suitable treatment as shown in Fig. 14.4.

The construction is that of an independent leaf, isolated from the existing wall. Where the existing wall is at least 100 mm thick of a masonry construction and plastered both sides, the independent leaf needs to be fitted to one face only. Where the wall is of any other form of construction an independent leaf is required to both faces.

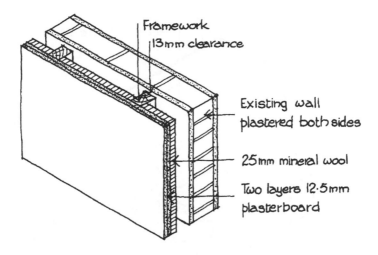

Fig. 14.4 Sound insulation of existing walls

An important feature of this type of sound insulation is that the independent leaf must, at all points, be completely isolated from the structural wall. The slightest contact renders the treatment ineffective.

The construction of the leaf is a framework, usually, but not necessarily, of suitably sized timbers spaced 13 mm away from the wall face. A 25 mm thick sound deadening quilt of mineral wool of a density of $10\,\text{kg/m}^3$ is hung between the framing members and the leaf is finished with two layers of 12.5 mm plasterboard fixed with the joints staggered. There should be a clearance of at least 25 mm between the plasterboard and the wall face.

The perimeter of the independent leaf should be sealed to the adjoining walls, floor and ceiling with tape or caulking.

An alternative construction for the independent leaf is to use a total thickness of 30 mm of plasterboard and no framing, with the mineral wool sandwiched (but not compressed) between the plasterboard and the existing wall.

Where the wall insulation is accompanied by one of the floor treatments described above, the independent leaf should run through to the existing structure and any floating floor or independent ceiling finished up to it.

14.2.5 Stairs – sound insulation

The underside of the stair, unless there is a cupboard formed below it, should be lined with an independent ceiling constructed as shown for a floor in detail A of Fig. 14.1.

If there is a cupboard under the stair this will assist in the sound insulation

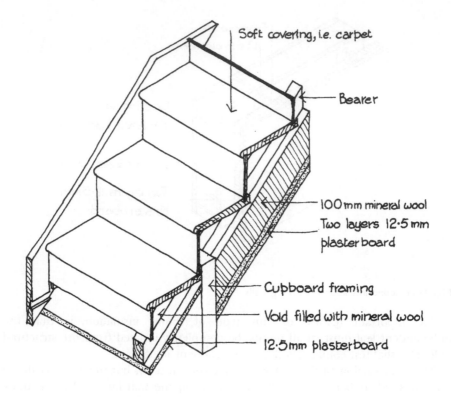

Soft covering, i.e. carpet

Bearer

100 mm mineral wool
Two layers 12·5 mm
plasterboard

Cupboard framing

Void filled with mineral wool

12·5 mm plasterboard

Fig. 14.5 Sound insulation of existing stairs

if the walls incorporate or are built from two layers of 12.5 mm plasterboard and the cupboard is provided with a small, heavy, well-fitted door. With this cupboard construction, the underside of the stair should be lined with 12.5 mm plasterboard and the space above filled with mineral wool (see Fig. 14.5).

14.2.6   Piped services

Piped services passing through a floor separating flats create both a potential fire hazard and nullify any sound insulation treatment. For these reasons, they should be enclosed both above and below the floor by a duct constructed as set out in Section 5.5 of Chapter 5, with the additional requirement that at no point should the duct construction rest on any floating floor. It should either stop short by 3 mm and this gap be filled with a suitable flexible fire stopping or, which is probably better, it should be carried down to the structural floor with the floating floor stopped 10 mm short of the duct face and the gap filled with a resilient seal.

14.3   Loft conversions

In the majority of cases, references in the Approved Documents to loft conversions are with regard to a relaxation of the Building Regulation requirements. In all other aspects, not specifically mentioned, of the conversion of a roof space into a habitable room, the work must comply with the requirements in exactly the same way as any other new building. The sections which follow show the relaxations permitted.

14.3.1   Fire resistance of floors

The standard requirement for a house which is to be altered in such a way that another storey is added is that the floor , whether it is old or new, should have a fire resistance of 30 minutes. This means that for that period of time it will not collapse, it will not allow the fire to break through and it will resist the transference of excessive heat but, provided that the conditions listed below are fulfilled, this resistance can be reduced to a 'modified 30 minute' standard. Modified 30 minutes means that the floor will still not collapse within that time but its resistance period for the fire breaking through and the excessive transfer of heat is reduced to 15 minutes.

The conditions attaching to this relaxation are:

1   only one storey is being added;
2   there will not be more than two rooms in the extra storey;
3   the total area of the new storey does not exceed $50 \, m^2$;
4   the floor only separates rooms, not circulation spaces;
5   the means of escape described in Section 14.3.3 and the enclosure of the staircase described in Section 14.3.7 are provided.

14.3.2   Rooflights – fire resistance

Many loft conversions obtain their natural light and ventilation by the use of a rooflight. If this is in the slope of the roof facing the boundary it could cause a fire hazard to any adjoining property. In this case the requirements of the Regulations and the details set out in Approved Document B4 must be observed. These requirements impose limits on the size and spacing of the rooflights and on the distance from the boundary. Rooflights glazed with a rigid thermoplastic sheet made from either polycarbonate or unplasticized PVC, or glazed with unwired glass are acceptable without any restrictions, provided that the thermoplastic materials achieve a Class 1 rating in a spread of flame test and that the unwired glass is at least 4 mm thick.

Rooflights using plastic materials such as single or multi-skinned poly-carbonate sheet less than 3 mm thick and PVC sheet which do not have a spread of flame rating better than Class 3 should have a maximum area of

$5\,m^2$, they should be spaced 3 m apart and be at least 6 m from the boundary. The maximum area requirement can be applied as a total to a group of rooflights located closer together than 3 m.

14.3.3 Rooflights and dormers as a means of escape

The room in the loft conversion should have either a dormer window or a rooflight which complies with the dimensions shown on Fig. 15.1 of Chapter 15. A door to a roof terrace is also acceptable as a means of escape.

The minimum size of opening and the height of the sill is to allow a person to get out through the window and the maximum distance from the eaves of the roof to the opening is to allow for rescue by ladder. As it is assumed that the secondary means of escape would be by ladder it is necessary for there to be a reasonable pedestrian access to a place where the necessary ladder can be set up.

The assumption of rescue by ladder is contrary to the general principle followed in the Approved Documents that a building occupant should be able to escape without outside assistance but, in the context of loft conversions, it is considered to be reasonable.

If there are two rooms in the loft, each should have an escape window if possible but, if not, a single escape window is acceptable provided that:

1 both rooms have their own access to the stair;
2 a communicating door is provided between the rooms so that an occupant of one room can gain access to the escape window in the other without having to pass through the stair enclosure.

14.3.4 Stairs – headroom

The standard requirement for the headroom over a stair is at least 2 m across the whole width. In a loft conversion it is considered that the headroom is adequate if it measures 1.9 m in the centre, reducing to 1.8 m at the side of the stair (see Fig. 14.6).

14.3.5 Stairs – alternating tread stairs

The alternating tread stair, sometimes referred to as a paddle stair, is a space saving arrangement in which part of the tread is cut away at alternate ends of each tread, allowing the flight to be fixed at a much steeper angle (see Fig. 14.7). The person using it ascends by stepping on to the wide portion of each tread alternately and he or she needs to be familiar with it to use it safely.

This form of stair should only be used as a straight flight, or flights, in a loft conversion leading to one habitable room only, plus, if so planned, a bathroom

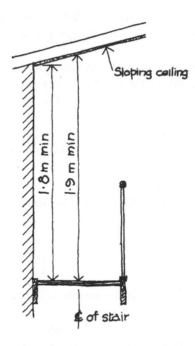

Fig. 14.6 Reduced stair headroom in loft conversions

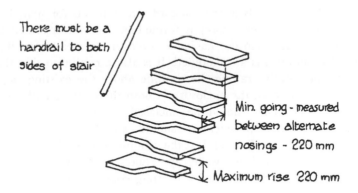

There must be a handrail to both sides of stair

Min. going - measured between alternate nosings - 220 mm

Maximum rise 220 mm

Twice the rise plus the going should be between 550 and 700 mm

Fig. 14.7 Alternating stair tread

or WC (which must not be the only WC in the dwelling). Nor should it be used if there is room to fit a stair designed in accordance with the principles set out in Chapter 9.

The going is measured from the nosing of the wider part of one tread to the nosing of the wider part of the tread next but one above and should not be less than 220 mm. The rise is measured in the normal manner, from the top surface of one tread to the top surface of the next tread above and should not be more than 220 mm. As for a normal stair, the sum of twice the rise plus the going should be between 550 mm and 700 mm. There should be a handrail fixed to both sides of the stair.

The treads should have a non-slip surface and the rise from one tread to the next should be designed so that a 100 mm sphere cannot be passed through the flight.

14.3.6 Ladders

A fixed ladder can be used as the means of access to a loft conversion but it should be to one room only and only then if there is not the room to provide a normal stair. There must be a fixed handrail to both sides of the ladder. Retractable ladders are not acceptable as a means of escape.

14.3.7 Stair enclosures

Chapter 15 gives details of the standard requirements for stairs in three storey houses. Where the third storey is formed by the conversion of the roof and there are no more than two habitable rooms in an area of less than 50 m^2 the standard requirements are modified. It is also a condition that the conversion works do not involve raising the roofline above the existing ridge level. The modifications concern the enclosure of both the existing and the new stairs.

The existing stairs should be enclosed by fire resisting walls giving 30 minutes' protection. This enclosure should extend from the ground floor up to the ceiling of the first floor and, at the ground floor level, lead either directly to a final exit or to at least two escape routes giving access to final exits and separated from each other by fire resisting construction in the manner shown in Fig. 15.2.

The new stairs should be located in the same enclosure extended up into the roof space, or enclosed in their own fire resisting stairwell, separated from the enclosure to the existing stairs and the ground and first floor rooms but opening into the existing stair enclosure at first floor level.

All habitable room doors which open into the stair enclosure should be fitted with a self-closing device. The existing doors, except for any glazed panels in them, can remain as they are but any new doors should be of a fire resisting construction.

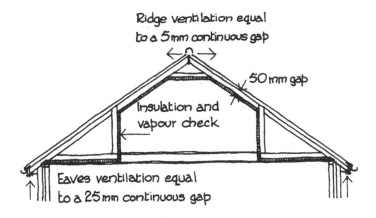

Fig. 14.8 Ventilation of roof space

All glazing within the enclosure to the existing stairs, except that to a bathroom or WC compartment, should be fire resisting. This includes any door panels, whether or not the door itself needs to be fire resisting.

14.3.8 Insulation and condensation

Where a room is created in a roof space, the cross-ventilation, which may have been provided when it was constructed (if the building is of recent age) or afforded by leakage in an older roof, is interrupted. This loss plus the greater degree of warming which will inevitably take place when the room is used, can give rise to a very real risk of condensation occurring. To counteract this, there should be eaves ventilation equivalent to a continuous gap 25 mm wide (if it has not already been provided) and ridge ventilators equivalent to a continuous gap 5 mm wide. There must also be an air space of at least 50 mm between the room enclosure, where it follows the roof slope, and the soffit of the roof, as shown in Fig. 14.8.

Lining the room enclosure with a polythene vapour check, placed on the warm side of all the insulation, will restrict the amount of water vapour entering the roof voids and thereby reduce the condensation problem, but vapour checks on their own cannot be relied upon to give a complete alternative to ventilation.

The Building Research Establishment has produced a Report with the title *Thermal insulation: avoiding the risks.* This is enclosed with the Approved Documents (although it is not, of itself, an Approved Document) and the recommendations it contains should be followed as good practice.

14.4   Conservatories

A Conservatory is defined, in Approved Document L, as a room with not less than three-quarters of its roof and not less than half its walls made of a translucent material.

If there is no separation between the conservatory and the rest of the house it must be treated as an integral part of the dwelling, subject to all the relevant Building Regulation requirements.

A separated conservatory is one that has been built against the exterior of the property so that there is, between it and the interior of the dwelling, a wall with a U value of at least 0.6 W/m^{2}K and any windows and doors have the same standard of U value and draught stripping as elsewhere in the dwelling. In this case the conservatory does not need to be heated, indeed Approved Document L points out that energy savings can be made if it is not heated. If, however, fixed heating equipment is fitted, it should have its own, separate, temperature and on-off controls.

14.4.1 Glazing

The same rules in connection with safety apply to the glazing of a Conservatory as apply to glazing in internal partitions and in windows and doors. These are set out in detail in Clauses 7.6 and 8.4. Briefly, the requirement is that precautions should be taken to prevent injury following accidental breakage. There are four ways in which this can be achieved:
- If it breaks it should do so safely, by this is meant that the impact results in either a small, clear opening only with a limit to the size of the detached particles, or disintegration into small detached particles that are neither sharp nor pointed.
- It is sufficiently robust to withstand the impact. Toughened annealed or laminated glass would be suitable for this.
- It is glazed in small panes that do not exceed 250 mm in width or 0.5 m^2 in area.
- The glazing is fitted with robust permanent protection to 800 mm above the floor level through which a sphere of 75 mm diameter cannot pass.

Chapter 15

MEANS OF ESCAPE AND FIRE FIGHTING FACILITIES

| 15.1 The Building Regulations applicable |

B1 Means of escape
B3 Internal fire spread (structure)
B5 Access and facilities for the fire service
K2 Protection from falling

B1 is the main area of legislation on this subject and states that the building must be designed and constructed so that there are means of escape, in case of fire, from the building to a place of safety outside the building, capable of being safely and effectively used at all material times; B3, as part of its general requirements for fire resisting structure, sets out the standards for protected shafts containing a staircase used as a means of escape; B5 requires that the building be designed and constructed so as to provide facilities to assist fire fighters in the protection of life, and provision made within the site to enable fire appliances to gain access; K2 is involved where a means of escape across a flat roof requires guarding.

| 15.2 Other legislation |

The relation between the Building Regulations and the Fire Precautions Act is that the fire authority cannot impose further conditions in relation to means of escape if the local authority is satisfied that the proposed works comply with the requirements of the Building Regulations, unless there is some aspect which is outside the details required for the purpose of Building Regulation consent. It should be noted, however, that there are some things which can be required under the Fire Precautions Act which do not appear in the Building Regulations. An example of this is the provision of fire fighting equipment for use by the occupants.

The Housing Act 1985 obliges the local authority to require means of escape from houses in multiple occupation. This means that the people living in the house are not of a single household. Compliance with the guidance given in the Approved Documents to the Building Regulations would be taken as an acceptable standard of safety for this purpose.

Not only must adequate provisions be made in the first instance, they must also be properly maintained and the Approved Documents assume that the building will be properly managed. Failure to carry out such management may result in the prosecution of the building's owner or occupier under legislation such as the Fire Precautions Act or the Health and Safety at Work Act, coupled with possible prohibition of the use of the premises.

15.3 General principles

Any means of escape should be designed to take into account the form of the building, the activities of the occupants, the likelihood of fire and the potential source and spread of a fire. It is generally accepted that blocks of flats and similar residential buildings occupied by a number of people or households represent a high fire risk by reason of the segregation of the occupancies and the nature and variety of the activities pursued. Careful observance of the guidance given in the Approved Documents on the subjects of fire resisting structures or linings and means of escape is, therefore, essential.

It is also generally accepted that normally a fire only starts in one place in a building and initially it creates a hazard only in the small area where it starts. Subsequently it is likely to spread through the building, often along the circulation routes. The chance of it starting in a circulation area is small provided that the combustible content of such areas is restricted.

There is also less risk that a fire will start in the structure of the building than in items such as furnishings which are not subject to the Building Regulations.

Smoke and noxious gases, not the flames, are the primary danger from a fire. Not only do they obscure the way to escape routes and exits, they may also actually cause casualties. Measures designed to provide safe means of escape from the building must, therefore, take into account the limitation of the rapid spread of smoke and gases.

In providing fire protection of any kind it should be borne in mind that any measures taken which interfere with the day-to-day use of the premises and cause inconvenience to the occupants are likely to be unreliable due to action being taken to eliminate the inconvenience which also eliminates the effectiveness of the fire safety provision.

15.3.1 Definitions

Occupant capacity Either the maximum number of people the room or storey is designed to hold, or the number found by dividing the floor area (excluding stairs, lifts and sanitary accommodation) by the floor space factor given below:

Dining room or lounge	$1.0 \, m^2/person$
Kitchen	$7.0 \, m^2/person$
Bedroom	$8.0 \, m^2/person$
Bed-sitting room	$10.0 \, m^2/person$

Travel distance The length of the shortest route which, if it includes a stair, is measured along the pitch line down the centre of the steps.

Width of a door The dimension across the door – not the clear width between the stops.

Width of an escape route The dimension at 1.5 m above the floor or stair pitch line, ignoring any handrails.

Width of a stair The clear distance between the walls or balustrades. Strings protruding less than 30 mm and handrails protruding less than 100 mm may be ignored.

15.3.2 Criteria for means of escape

There are two basic principles for the design of means of escape. Firstly, in most situations, there should be an alternative means of escape. Secondly, where direct escape to the open air clear of the fire cannot be achieved, it should be possible to reach a place of relative safety, such as a protected stairway, within a reasonable travel distance.

The following are not acceptable as means of escape:

1 Lifts, except for purpose designed evacuation lifts intended for use by disabled people.
2 Portable ladders and throw-out ladders.
3 Manipulative apparatus and appliances such as fold-down ladders.

15.3.3 Unprotected and protected escape routes

The first part of any escape route is usually unprotected and consists of the distance travelled through the accommodation and along any corridors . This should be kept as short as practicable so that people do not have to travel very far while exposed to the immediate danger of fire and smoke.

At the end of the unprotected escape route should be either the ultimate place of safety which is the open air clear of the effects of the fire, or a protected corridor or, more commonly, a protected stairway. The length of a protected corridor should be limited because the structure will not provide adequate protection indefinitely.

Protected stairways are designed to provide what is, in effect, a 'fire sterile' area which leads to the ultimate place of safety outside the building. Once they have reached a protected stairway, anybody should be able to consider themselves safe from the immediate dangers of the fire and can then proceed to a place of safety at their own pace. To achieve this ideal, flames, smoke and gases must be excluded from the protected stairway as far as is reasonable by the provision of a fire resisting structure, or a smoke control system or both.

Not all stairways must be protected, 'accommodation stairs' can be installed for regular daily use but their role in terms of fire escape is only very limited.

An escape route which crosses a balcony or flat roof should be guarded to prevent anybody falling. The means of guarding is the same as that to be provided for landings to stairs or ramps and should comply with the requirements of Regulation K as set out in Chapter 9 of this book.

15.4 General provisions applicable to all houses, flats and maisonettes

The feasibility of being able to escape from within a house, bungalow, flat or maisonette is dependent on very similar factors related to the design of the plan of the dwelling, its details of construction and provisions made within it. These are described in the following sections. With houses and bungalows, once the occupants have escaped from within the dwelling they are usually outside the building and safe from any danger. With flats and maisonettes, the escape from within the dwelling is only the first part of the escape route, the second being along corridors common to all the dwellings and down staircases to the outside and is dealt with separately in Section 15.6.

In providing fire protection of any kind it is necessary to realize, and make due allowance for the fact, that if any of the protective measures interferes with the convenience of the occupants it is likely that it will be rendered ineffective by unauthorized actions.

15.4.1 Inner rooms

Any room that does not connect directly with a corridor or hall is termed in Approved Document B1 an 'inner room' and the room through which it is necessary to pass to get to it is the 'access room'. It is obvious that a fire in the access room would immediately trap anybody in the inner room and, for this reason, an inner room in a plan is only acceptable if it is one of the following:

1 a kitchen;
2 a laundry or utility room;
3 a dressing room;
4 a bathroom, shower room or WC;
5 any other room in the basement or on the ground floor or first floor which has a window or door suitable as a means of escape.

15.4.2 Basements

If a basement contains a habitable room there should be a means of escape such as a window or external door which is an alternative to the stairs, to avoid it being necessary for someone escaping from a fire in the basement having to go up into the layer of smoke which has risen and collected on the ground floor. If the basement accommodation is not separated from the rest of the house, the rooms on the upper floors are, effectively, inner rooms and are subject to the limitations set out above.

15.4.3 Windows and external doors for escape

If a window or external door is required to afford a means of escape from a fire it must comply with the conditions listed below and as shown on Fig. 15.1.

1 It should have an opening at least 850 mm high and 500 mm wide.
2 If it is a window the sill should be between 800 and 1100 mm above the floor except for a rooflight where the sill may be 600 mm above the floor.
3 A dormer or rooflight should be positioned with its sill no more than 1.7 m up the roof slope from the eaves.
4 If the escape is to a courtyard or back garden with no way out except through another building, the distance from the house to the bottom boundary should be not less than the height of the building.

15.4.4 Smoke alarms

In a very large house, where the distance from any part of one room to the furthest part of another on the same floor is more than 30 m a system of smoke detectors and alarms connected to a central control and indicator unit is required.

In most houses, flats or maisonettes, the installation of a self-contained smoke alarm is considered satisfactory. The alarm should be either mains operated or a mains unit with back-up battery supply (battery only fire alarms are not acceptable), installed to comply with the following:

1 Within 7 m of the door to a kitchen or living room and 3 m from a bedroom door in the circulation areas.

Ceiling fixing alarms should, preferably, be placed centrally and at least 300 mm from any light fittings.

Wall fixing alarms should be fitted between 150 and 300 mm below the ceiling.
2 They should not be fitted:
   (a) next to or directly above a heater or air conditioning outlet;
   (b) in a bathroom, shower room, cooking area or garage or anywhere where steam, condensation or fumes could give false alarms;
   (c) anywhere that gets very hot (such as a boiler room) or very cold (such as an unheated porch);

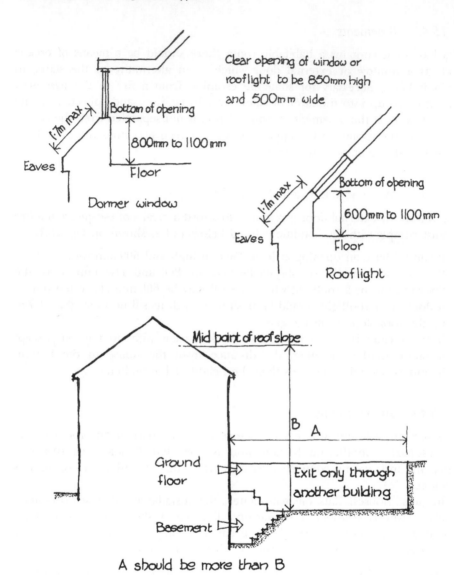

Clear opening of window or rooflight to be 850mm high and 500mm wide

Bottom of opening

1·7m max

800mm to 1100mm

Eaves

Bottom of opening

Floor

Dormer window

1·7m max

Bottom of opening

600mm to 1100mm

Eaves

Floor

Rooflight

Mid point of roof slope

B A

Ground floor

Exit only through another building

Basement

A should be more than B

Fig. 15.1 Escape criteria

(d) on surfaces that get much warmer or colder than their surroundings (the temperature difference may cause air currents which divert the smoke away from the alarm).

3 A corridor over 15 m long should have more than one smoke alarm and they should be interconnected so that the detection of smoke by one sets off all of them.

Note: this does not apply to a corridor in a block of flats or maisonettes common to all the dwellings.

4 In a house of more than one storey and a maisonette there should be smoke alarms in each storey and all alarms within each dwelling should be interconnected.

5 There should be a separate electrical circuit for the alarms from the distribution board of each dwelling, fitted with its own fuse. The system may operate at low voltage through a mains transformer. The wiring is to conform to the IEE Wiring Regulations but the cable used does not need to have any special fire survival properties.

15.4.5 Ducted warm air heating systems

The Approved Document B1 states that with this form of heating, precautions need to be taken to prevent the spread of smoke or fire into a protected stairway and refers to BS 5588: Part 1 as a source of guidance.

The British Standard sets out five recommendations:

1 Transfer grilles should not be fitted in any wall, floor or ceiling of a protected stairway.

2 Any ductwork passing through a protected stairway should have all the joints between the ducting and the stairway enclosure fire stopped.

3 Where ductwork is used to convey warm air into a protected stairway, ductwork is also required for the return air from the stairway back to the heater.

4 Warm air and return air grilles or registers should be positioned no higher than 450 mm above the floor level.

5 A room thermostat should be positioned in the living room at a height between 1370 and 1830 mm and its maximum setting should not exceed 27°C.

15.5 Houses and bungalows

Normal bungalows and houses rising to no more than two storeys present very little difficulty with regard to means of escape provided that all habitable rooms either open directly into a hall or corridor leading to an entrance or have a window or external door through which escape can be made and provision has been made to give early warning of a fire.

As the height increases the hazards also increase and protection to an internal stairway is required where the upper storey floor is more than 4.5 m above ground level. In a house with a floor 7.5 m or more above the ground level the risk of someone being trapped has grown to the point where he or she may not be able to get down the stairs in time and an alternative escape route is necessary.

15.5.1 Three storey houses

In any house where there is a floor over 4.5 m above ground level – as is likely in a three storey property – the chance of escape through a window is reduced and it is necessary to provide an alternative way of reaching safety. The Approved Document B1 gives two possibilities: either a protected stairway serving the upper storeys or the top floor can be separated from the lower storeys by appropriate fire resisting constructions and be provided with its own means of escape. It should be noted that BS 5588: *Fire precautions in the design, construction and use of buildings*, Part 1: *Code of practice for residential buildings*, states that lowering lines and portable, retractable and throw-out ladders are not considered suitable for means of escape purposes. Vertical or raking ladders permanently fixed in the position required will only be suitable in exceptional circumstances.

If a protected stairway is to be provided it should finish at ground floor level either in a hall leading directly outside or in a position with at least two escape routes leading to the outside (see Fig. 15.2).

15.5.2 Houses of four or more storeys

In houses of this height where there is more than one floor that is over 4.5 m above ground level additional precautions need to be taken and for these the Approved Document B1 refers to the guidance given in clause 4.4 of BS 5588: Part 1.

The British Standard sets out the following recommendations:

1 A habitable room should not be an inner room unless it is less than 4.5 m above ground level and is provided with a door or window suitable as a means of escape or rescue.
2 All floors, including the ground floor, should be connected to a stair that is constructed as a protected stairway and delivers either directly to a final exit or to at least two escape routes leading to alternative final exits.

If the building is of such a height that a floor or floors are more than 7.5 m above ground level, there should be an alternative escape route from each of these upper storeys. Where the access to this alternative escape route is by way of a protected stairway – either to an upper floor or to a way out on the same floor – this protected stairway should be separated from the lower storeys by a fire resisting construction at or about the 7.5 m level.

15.6 Flats and maisonettes

Where the block of flats or maisonettes is of a similar size to a house, i.e. a ground floor and first floor with, possibly, a basement in addition, the means of escape requirements set out in the Approved Document are very similar and require that:

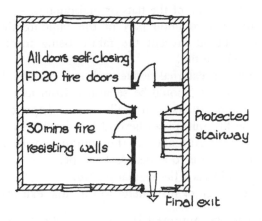

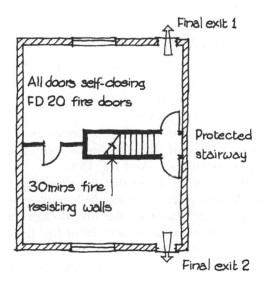

Fig. 15.2 Alternative final exits

1 each habitable room opens either directly into a hall or corridor leading to the outside or that it has a window or external door suitable as a means of escape or rescue,
2 smoke detectors and alarms are provided to give early warning of a fire.

As more floors are added and the height of the block increases, the hazards of escaping through a window also increase and, therefore, more complex escape provisions need to be made.

The guidance given in Section 2 of Approved Document B1 is based on the following assumptions:

1 The fire is generally in one of the flats or maisonettes.
2 If the fire occurs in a common part of the building the materials and construction used should prevent the fabric being involved beyond the immediate vicinity of the fire.
3 There is no reliance on external rescue such as a portable ladder.
4 As each flat or maisonette should be separated from its neighbours by fire resisting compartment floors and walls (see Chapters 5 and 7) the probability that the fire will spread beyond the dwelling where it originated is low and simultaneous evacuation of the building not likely to be needed.
5 There are two distinct parts to a means of escape:
    (a) escape from within each flat or maisonette;
    (b) escape from each flat or maisonette to the final exit from the building.

The guidance given in the Approved Document is also applicable to any flat or maisonette considered to be in multiple occupation.

15.6.1 Internal planning of flats in blocks of four storeys or more

There are three ways that a flat with a floor more than 4.5 m above ground level can be planned, all of which have one common element – no habitable room should be an inner room unless it has an external door or window suitable as a means of escape or rescue.

The first planning method is to provide a protected entrance hall to which all habitable rooms are connected and in which no room door is further than 9 m from the flat entrance (see plan A of Fig. 15.3).

The second way is to plan the flat so that the travel distance from the flat entrance door to the furthest point of any habitable room is not more than 9 m and the cooking facilities are placed so that they do not present a potential hazard to the escape route (see plans B1 and B2 of Fig. 15.3).

The third approach is to provide an alternative exit from the flat. This exit should be located in the part of the flat containing the bedrooms and these should be separated from the rest of the flat by fire resisting construction and self-closing fire doors (see plan C in Fig. 15.3).

15.6.2 Internal planning of maisonettes in blocks of four storeys or more

There are two acceptable ways to plan a maisonette which does not have its own entrance at ground level and has one of its floors more than 4.5 m above ground level.

The first is to provide an alternative exit from all habitable rooms which are not on the entrance hall level of the maisonette (see plan A in Fig. 15.4).

The second way is to provide one alternative exit on the upper floor and fire

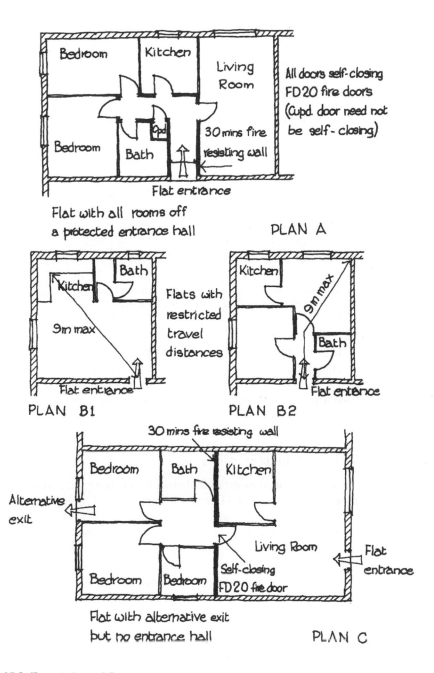

Fig. 15.3 Typical plans of flats

resisting enclosures to the entrance hall and upper floor landing (see plan B of Fig. 15.4).

15.6.3 Alternative exits

To be effective an alternative exit from a flat or maisonette should comply with the following:

1 Be remote from the main entrance door of the dwelling.
2 Lead to a final exit or common stair by way of one of the following:
   (a) a door on to an access corridor or common balcony;
   (b) an internal private stair leading to an access corridor or common balcony at another level;
   (c) a door on to an external stair;
   (d) a door on to an escape route over a flat roof.

15.7 Means of escape from flats and maisonettes

Having reached the exit from the flat or maisonette in safety, the escaping occupant has then to make his or her way to a final exit and the ultimate safety of the outside of the building. The way this is achieved depends principally on the height of the building (or its number of floors) but, in addition, on whether the dwellings are reached via an internal stairway or have balcony or deck access.

15.7.1 Buildings up to three storeys

If the block containing the flats or maisonettes is no more than three storeys high and the top floor is not over 11 m above ground level, a single stairway in a protected shaft is considered adequate as an escape route. This single stairway should be reached through a protected hall or lobby, as shown in Fig. 15.5, and should not connect to a covered car park unless it is open sided.

The travel distance from the flat or maisonette entrance to the door leading to the stairway should not exceed 4.5 m unless the hall or lobby has an automatically opening vent (to remove any smoke) in which case the travel distance can be increased to 7.5 m.

15.7.2 Buildings over three storeys

The general principle applied to taller buildings is that it must be possible for any persons confronted by the effects of a fire to turn away from it and make their escape by an alternative route. In all cases, the maximum travel distance from the flat or maisonette entrance to the door to the protected shaft containing the stairs should not exceed 30 m (see Fig. 15.6).

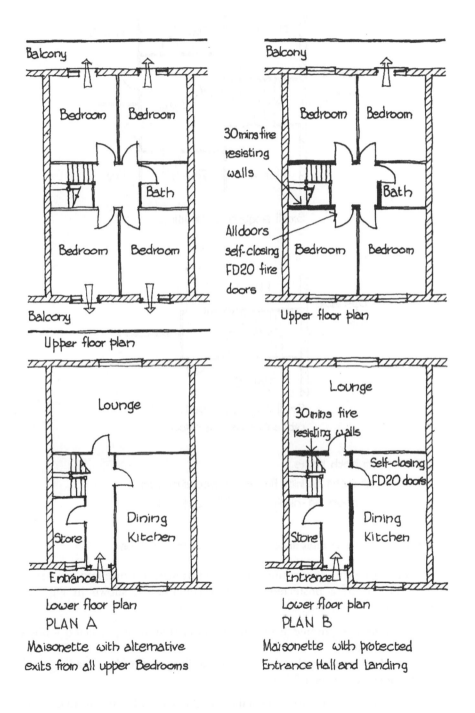

Fig. 15.4 Typical maisonette plans

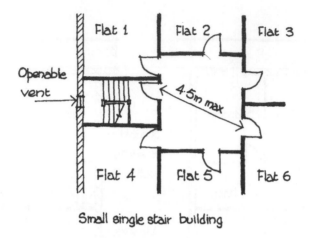

Small single stair building

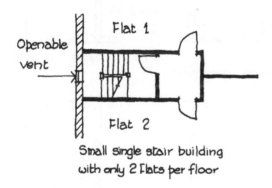

Small single stair building
with only 2 Flats per floor

In both plans:
All walls 30 mins fire resisting construction
All doors self-closing FD30S fire doors

Fig. 15.5 Escape routes in small single stair buildings up to four storeys

However, a single escape route from the entrance to a dwelling via a single common stairway in a protected shaft is acceptable, as shown in Fig. 15.6, provided that:

1 every dwelling is separated from the common stair by a protected lobby or common corridor;
2 the travel distance from the door of the dwelling to the door leading to the stairway does not exceed 7.5 m.

A single escape route is also acceptable in the case of flats or maisonettes located in a dead end part of a common corridor served by two or more

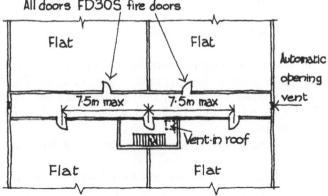

Doors to Flats: FD30S fire doors
Doors across Corridor: FD20S fire doors

Flat 13   Flat 11   Flat 5   Flat 3   Flat 1   Automatic opening vent

30m max   7·5m max

Flat 8   Flat 4   Flat 2   Operable vent

Plan repeated opposite hand

Plan of Flats block with two stairways

All doors FD30S fire doors

Flat   Flat   Automatic opening vent

7·5m max   7·5m max   Vent in roof

Flat   Flat

Plan of tower block with single internal stairway

Fig. 15.6 Escape route travel distances – buildings over four storeys

common stairs provided that the travel distance from the dwelling entrance to the stairs does not exceed 7.5 m.

15.7.3 Flats or maisonettes with balcony or deck access

For this situation, Approved Document B1 again refers to BS 5588: Part 1 which recommends that the balcony connects to more than one stair in a protected shaft or, if it is a single, central access stair, the walls between the dwelling and the balcony should be of fire resisting construction up to a height of 1.1 m above the deck level and any doors in these walls should be self-closing FD 20 fire doors (see Chapter 8 for the meaning of FD 20). BS 5588 also points out the hazards connected with wide balconies in which there is a danger that the width can make it possible for the balcony to become smoke-logged even though it is open on one side and the greater width is a temptation to erect 'external' store rooms or use the space to create other fire risks. Accordingly, the British Standard recommends that, where the balcony is more than 2 m wide, the soffit above it should be fitted with downstands at 90° to the face of the building, on the line of separation between dwellings, projecting 0.3 m to 0.6 m below any other beam or downstand parallel to the face of the building to control the spread of smoke and no store or other fire risk should be erected on a balcony.

15.7.4 Common escape routes

Escape routes to be used by more than one dwelling should be planned so that people do not have to pass through one stairway enclosure to reach another, the length of travel distance should be limited to the dimensions shown in the plans above.

The walls and floors enclosing the corridors should be compartment walls and floors with a 30 minute fire resistance (see Chapter 7). Glazing is permitted in these enclosing walls and in any fire doors but only of the limited extent set out in Chapter 8 and Table 8.4 which is reproduced here for convenience as Table 15.1.

All escape routes should have a clear headroom of not less than 2 m at all points.

The floor of an escape route should be covered with a flooring which will not become slippery if it gets wet. Any ramps in the escape route should be as easy as possible and no steeper than a 1 in 12 gradient. In addition, ramps should be designed in accordance with the guidance given in Approved Document K and as described in Chapter 9.

Approved Document B1 points out that, despite the precautions taken in compliance with recommended practice, it is possible that smoke will still get into common corridors or lobbies, if only because the occupant of the dwelling on fire has to open their door to escape. There should, therefore, be the following means of ventilation provided in these areas:

1 In a single stair building, other than a low rise building as shown in Fig. 15.5, and in any dead-end parts of buildings with more than one stair, the common corridor should be ventilated by an automatic ventilating device, located in the area to be ventilated, with a free area of not less than 1.5 m^2 and fitted with a manual override.

Table 15.1 Limitations on uninsulated glazing on escape routes

Position of glazed element	Maximum total glazed area in parts of a building with access to a single stairway	
	Walls	Door leaf
Within a protected stairway in a single family house	Fixed fanlight only	Unlimited
Within a protected entrance hall or landing of a flat or maisonette	Fixed fanlight only	Unlimited above 1.1 m from the floor
Between a protected stairway and the accommodation or unprotected corridor	Nil	25% of the door area
Between the accommodation and a protected lobby or protected corridor forming a dead end	Unlimited above 1.1 m from the floor	Unlimited above 0.1 m from the floor
Between a protected stairway and a protected lobby	Unlimited above 1.1 m from the floor	Unlimited above 0.1 m from the floor

(*Source:* Based on Table A4 of Approved Document B)

2 In buildings with more than one stair, common corridors should extend to the external wall at both ends and be fitted with openable ventilators, with a free area of not less than 1.5 m^2, for fire service use. These ventilators may be operated automatically.

Store rooms and other ancillary accommodation should not be located in, or entered from any protected lobby or corridor which forms part of an escape route.

15.7.5 Doors on escape routes

Doors on escape routes should be fire resisting and have the performances set out in Table 8.3 of Chapter 8 which, for convenience is reproduced here as Table 15.2.

Fire doors should be provided to subdivide corridors where they connect two or more exits and to separate any dead ends from the rest of the circulation system.

As any fire door should be shut to be effective, anybody attempting to escape is inevitably delayed to some extent by the need to open the door. This delay must be kept to the minimum and Section 5 of Approved Document B1 gives the following guidance:

1 It is preferable that doors are not fitted with a lock, latch or bolt but if a fastening is required it should be simple, readily operated from the side

Table 15.2 Provisions for fire doors

Position of door	Minimum fire resistance in terms of integrity (minutes)
In a compartment wall separating buildings	As for the wall in which it is fitted but not less than 60
In a compartment wall:	
between a flat or maisonette and a common area	FD30S
enclosing a protected shaft forming a stairway to flats or maisonettes	FD30S
not described above	As for the wall in which it is fitted*
Forming part of the enclosures of:	
a protected stairway except in a single family house	FD30S
a lift shaft or service shaft	FD30
Forming part of the enclosures of:	
a protected lobby or protected corridor to a stairway	FD30S
a protected corridor other than to a stairway	FD20S
Affording access to an external escape route	FD30
Subdividing:	
corridors connecting alternative exits	FD20S
dead end parts of corridors from the rest of the corridors	FD20S
a dwellinghouse and a garage	FD30
Forming part of the enclosures to:	
a protected stairway in a single family house	FD20
a protected entrance hall or protected landing in a flat or maisonette	FD20
In any other fire resisting construction not described above	FD20

(*Source:* Based on Table B1 of Approved Document B)

* If the door is used for escape the suffix S should be added.

approached by anybody escaping, without either the use of a key or having to manipulate more than one mechanism.

2 Where a self-closing device causes inconvenience the door may be held open by either a door closure delay device or an automatic release mechanism, provided that the door can be closed manually and does not lead to an escape stair or a fire fighting stair.

It is preferable that the direction of opening of a fire door should be the same as the direction of escape, if practicable. Where the number of persons likely to use the door exceeds 50 it must always open in the direction of escape.

It should open to not less than 90° and should not reduce the effective width of any escape route across a landing. If the door opens towards a corridor or a stairway it must be recessed sufficiently to prevent the door swing encroaching on the corridor or stairway width. Vision panels are needed in fire doors where they subdivide corridors or where they swing both ways.

Revolving doors, automatic doors and turnstiles should not be placed across escape routes where they can obstruct anybody escaping unless they can be opened easily in an emergency or are arranged to fail safe into an open position. Alternatively, where such a door is required, a normal swing door should be provided alongside it for the purpose of escape.

Final exit doors should be not less in width than the escape route they serve and sited to ensure the rapid dispersal of people from the vicinity of the building. They should also be sited away from any risk of fire or smoke issuing from basement ventilators or openings to transformer chambers, refuse chambers, boiler rooms and the like. Direct access to a street, passageway or open space should be provided and the route should be well defined and guarded where necessary. It is important that a final exit door is apparent to a person who may need to use it, particularly if it is from a stairway which continues up or down past the door.

All common escape routes in residential premises should have adequate artificial lighting installed in accordance with the recommendations in BS 5266: *Emergency lighting, Part 1: Code of practice for the emergency lighting of premises other than cinemas and certain other specified premises used for entertainment.* Escape stairs should be adequately lit and these lights should be on a separate circuit from that supplying the escape routes with the wiring protected. A protected circuit should be wired with cable classified as CWZ in accordance with BS 6387, run only through those parts of the building where the fire risk is negligible and separated from any circuit provided for another purpose.

15.7.6 Escape route over flat roofs

Where there is to be more than one escape route, one of these may be by way of a flat roof provided that:

1 the roof is part of the same building;
2 the route across the roof leads to a storey escape;
3 the part of the roof forming the escape route, its supporting structure and any openings within 3 m have a fire resistance of 30 minutes in respect of load bearing capacity, integrity and insulation;
4 the route is adequately defined;
5 the route is guarded as set out in Approved Document K and Chapter 9. **D15.1**

15.7.7 Common stairs

A common stair should have a minimum width of 1 m and be designed in accordance with the guidance given in Approved Document K and Chapter 9.

It should be constructed with materials of limited combustibility in any of the following circumstances:

1 if it is the only stair serving the building unless it is a block of flats of two or three storeys only;
2 if it is within a basement storey;
3 if it serves any storey with a floor level more than 20 m above the ground;
4 if it is external, but see Section 15.6.9 for further information on external escape stairs.

Single steps should be avoided because of the danger they present and are only acceptable on the line of a doorway.

Helical and spiral stairs (a helical stair is open in the middle whereas a spiral stair has a central column) are acceptable as part of an escape route and should be designed in accordance with the recommendations in BS 5395: *Stairs, ladders and walkways*, Part 2 and Approved Document K (see Chapter 9). Fixed ladders are not really suitable as a means of escape but can be used where it is not practicable to provide a conventional stair, such as a means of access to a plant room which is not very often occupied. They should be made of non-combustible material.

Common stairs should be enclosed in a protected shaft with a fire resistance of 30 minutes in respect of its loadbearing capacity, its resistance to the fire breaking through (its integrity) and its ability to give insulation from the effects of the fire. The enclosure should discharge directly to a final exit or else to a protected corridor leading to a final exit. Where two protected stairways adjoin, the wall between them must be imperforate.

There should be nothing else in a protected stairway except a lift installed in accordance with Section 5 of Approved Document B1 – see Section 15.8 below – nor should there be any gas service pipes or meters in the stairway enclosure unless they are in accordance with the relevant Gas Regulations (Gas Safety Regulations 1972 and the Gas Safety (Installation and Use) Regulations 1984 as amended by the Gas Safety (Installation and Use) (Amendment) Regulations 1990).

The fire resistance of the external wall of a protected stairway does not need to have a high degree of fire resistance but, if it does not offer much resistance to fire it must be separated from any other unprotected areas in the building by at least 1.8 m as shown in Fig. 15.7.

If there is only one common stair in the building it should not also serve any covered car park, boiler room, fuel storage space, or other ancillary accommodation of similar fire risk. A common stair which does not form part of the only escape route can give access to such spaces provided that they are separated by a protected lobby.

15.7.8 Basement stairs

Special measure are required with basement stairs to prevent a fire in this level endangering the upper storeys.

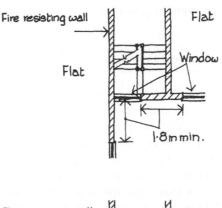

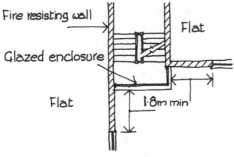

Fig. 15.7 Protection of stairway enclosure

If there is only one escape stairway in the building, it should not be extended down into the basement. In this case the basement must be served by its own stairway.

If there is more than one escape stairway in the building, one of them must terminate at ground floor level and the others can be carried down to the basement provided that there is a protected and ventilated lobby or corridor between the stairs and any accommodation in the basement.

15.7.9 External escape stairs

An alternative escape route may be via an external stair provided that:

1 it does not rise more than 6 m;
2 all doors giving access to the stair are fire resisting and self-closing (a fire resisting door is not required if it is the only exit to the top landing of the stair);
3 any part of the building within 1.8 m of the flights or landings, or 9 m below the stairs, should be fire resisting. The 1.8 m dimension may be

reduced to 1.1 m above the top level of the stair provided that it is not a stair up from a basement to ground level.

15.8 Flats in mixed use buildings

If the building is not more than three storeys high, a stairway may serve both residential and non-residential occupancies provided that the stairs are separated by protected lobbies from each occupancy at all levels.

In buildings over three storeys high, the stairs serving the flats should be separate from any other part of the building unless the flat is ancillary to the main use of the building. Where this is the case, the ancillary flat may be served by the same stairs that serve the rest of the accommodation but there must be protected lobbies separating the stairs from the non-residential parts in all the floors lower than the flat; there must be an alternative escape route from the flat and any automatic fire detection and alarm system in the main building areas must be extended into the flat.

15.9 Lifts

With the exception of those specially built to provide evacuation for disabled people, lifts should not be used as a means of escape. Lifts for the disabled should follow the recommendations of BS 5588: Part 8.

As well as not being suitable as a means of escape, lifts may prejudice escape by reason of their connecting one floor to another. To safeguard against this the following guidance is given in Approved Document B1:

1 Lift wells should be either within a protected stair well or be enclosed throughout its height by fire-resisting construction. A lift well connecting different compartments should form a protected shaft.
2 Unless it is within a protected stair well, the lift should only be approached through a protected lobby.
3 If the lift is in the enclosure of the only stair in the building or the one stair in the building which terminates at ground level, it should not be carried down to the basement.
4 Lift machine rooms should be located over the lift well if possible. If this is not practicable and the lift is within the enclosure to the only stairway in the building, the machine room must be located outside the enclosure.

15.10 Mechanical ventilation and air conditioning

Any system of mechanical ventilation should be designed so that the air is directed away from protected escape routes and exits or, if a fire occurs, the system or appropriate parts of it close down. Any systems which recirculate air should meet the relevant recommendations in BS 5588: Part 9.

Guidance on the design and installation of mechanical ventilation and air conditioning systems is contained in BS 5720 and on ventilation and air conditioning ductwork in BS 5588: Part 9.

15.11   Access and facilities for the fire services

Section 15 of Approved Document B1 gives guidance for the selection and design of facilities for the purpose of protecting life by assisting the fire service. The facilities covered are access for fire appliances, access for fire fighting personnel, the provision of fire mains within the building and the venting of smoke and heat from basements.

In the type of buildings generally within the scope of this book, the combination of personnel access by the normal means of escape and the ability to work from ladders and appliances on the perimeter will be adequate without special internal arrangements. If, however, the block of flats or maisonettes is more than 20 m high or contains a storey of more than 600 m^2 area at a height of more than 7.5 m above ground level, or it has two or more basement storeys exceeding 900 m^2, fire fighting shafts, possibly with fire fighting lifts, are required and the guidance given in Section 17 of Approved Document B1 should be followed.

15.11.1   Fire mains

The purpose of fire mains is to provide the fire fighters with points to which they can connect their hoses to fight a fire from within the building. They may be 'dry mains' in which case they are empty pipes which are connected at ground level to a hose from the fire service pumping appliance, or they may be 'wet mains' which are kept full of water supplied from tanks and pumps in the building. Wet mains should be provided with the facility to replenish the supply from a fire service pumping appliance.

Any buildings less than 60 m high can be fitted with either wet or dry mains, over this height it should always be a wet rising main.

Further guidance can be obtained from Sections 2 and 3 of BS 5306: Part 1 on the design and construction of fire mains. Fire mains are only required in buildings which have fire fighting shafts.

15.11.2   Vehicle access

Access is required to enable high reach appliances, such as turntable ladders and hydraulic platforms, to be used and for pumping appliances to supply water and equipment for fire fighting and rescue.

Table 15.3 Typical fire service vehicle access route specification

Appliance type	Minimum carrying capacity (tonnes)	Minimum dimensions of: (m)				
		Road between kerbs	Gateways	Turning circle between kerbs	Turning circle between walls	Clearance height
Pump	12.5	3.7	3.1	16.8	19.2	3.7
High reach	17.0	3.7	3.1	26.0	29.0	4.0

(*Source:* Based on Table 20 of Approved Document B1)

Note: Some Fire Authorities have appliances of greater weight or other sizes and the Building Control Authority may require a different specification.

Buildings without fire mains which are up to 2000 m² in area and with a top storey less than 9 m above ground level should be designed to allow vehicular access to within 45 m of any point of the area of the projected plan, or to 15 per cent of the perimeter, whichever is the easier to achieve. Buildings over this size not fitted with fire mains should provide access for fire service vehicles as set out in Table 19 of Section 17 of Approved Document B1.

The access route for the fire service vehicle may be a road or other route and should meet the criteria set out in Table 15.3. Where such access is provided, overhead obstructions such as cables or branches should be avoided within a zone 12.2 m out from the building and of a height equal to the height to the ridge. Any dead-end access roads should be provided with a turning circle based on the turning circle dimensions given in Table 15.3.

15.11.3 Venting of basements

The products of combustion tend to escape via the stairways which are to be used by the fire fighting personnel, making access, search, rescue and fire fighting difficult. The provision of smoke outlets – also called smoke vents – allows the smoke and heat to escape and can also be used to allow cooler air into the basement.

Where practicable, smoke outlets should be provided in each basement space. It may be that the building shape is such that there are basement areas which are not on the perimeter. In these situations it is acceptable that only the perimeter spaces are vented directly, the inner areas being vented indirectly by the fire fighters opening the connecting doors. If, however, the basement is compartmented, each compartment must be directly vented and not have to rely on opening doors in the compartment walls.

Smoke outlets are not required in a basement to a single family house or to one where the floor area is either less than 200 m^2 or not more than 3 m below ground level.

Naturally venting outlets should be sited either in the ceiling or at high level in the wall of the basement, evenly distributed around the perimeter of the building so as to discharge to the open air. Their total free area should be at least 2.5 per cent of the floor area of the space they serve and separate outlets should be provided anywhere where there is a special fire risk.

Generally the outlet can be covered by a panel, stallboard or pavement light which should be clearly indicated and which can easily be broken out. If the outlet has to be at a point which is not easily accessible it should only be covered by a non-combustible grille or louvre. The position of the smoke outlets must be carefully selected to avoid causing any difficulty in the use of a fire escape route from the building.

Basements fitted with an automatic sprinkler system which conforms to the recommendations of BS 5306: Part 2 can also have a mechanical smoke extract system which is activated by the sprinkler system. Alternatively, it may be activated by an automatic fire detection system conforming to BS 5389: Part 1 (at least L3 standard). The mechanical extract system should be capable of achieving at least ten air changes per hour and handling gases at temperatures up to 400°C for not less than one hour.

The outlet ducts and shafts should be enclosed by a non-combustible construction.

Appendix A

SAP ENERGY RATING CALCULATION WORKSHEET

(Based on Appendix G of Approved Document L to the Building Regulations 1995) **DA.1**

This worksheet is an abbreviated form of the worksheet given in Appendix G. **DA.2** It is only applicable to a limited range of houses or bungalows conforming to the descriptions given below. If the property under examination is not covered by all the specifications given, the full Appendix G worksheet should be used.

Note that the fuel costs and energy cost factor in Part 9 will be subject to periodic updating and the current values should be confirmed.

Specification:
The property covered by this worksheet must conform to the following specifications:

One or two storeys high.

Detached or semi-detached with separate garage building.

One bathroom and separate W.C.

Total floor area between 60 m² and 200 m².

External walls of bricks or blocks, insulated to a U value of 0.45 W/m²K.

Concrete ground floor insulated to a U value of 0.45 W/m²K.

Timber framed pitched and tiled roof insulated to 0.25 W/m²K.

Timber or PVC-U windows with normal 6 mm double glazing to a U value of 3.3 W/m²K.

Timber external doors half glazed with 6 mm double glazing to a U value of 3.1 W/m²K.

Draught lobbies provided to external doors.

Natural ventilation (except extract fans in bathrooms etc.).

Hot water radiators with thermostatic valves, heated by a gas boiler with balanced, fan assisted or open flue, or oil fired boiler controlled by a programmer and a room stat.

Standard 120 litre cylinder with 35 mm factory applied foam lagging and cylinder stat.

All primary pipework insulated.

Two heating zones: living areas (zone 1) and sleeping areas (zone 2).
Solid fuel open fire in lounge.
Gas price £4.26/GJ plus £38 standing charge, oil price £3.68/GJ, coal
£3.87/GJ.

Calculation of SAP Rating:

1. Sizes of the dwelling
*All plan dimensions are taken to the internal faces of the external walls and
include internal partitions. All heights are taken from finished ground floor level
to the highest ceiling and include any upper floor.*

Ground floor area: ☐ (1)

Total floor area: ☐ (2)

Total volume: ☐ (3)

2. Ventilation rate

Air changes per hour: $\dfrac{80}{(3)}$ ☐ (4)

Additional structural and window infiltration:

0.45 *for two storey or* 0.35 *for single storey* ☐ (5)

Air change rate: $[(4) + (5)] \times 1.85$ ☐ (6)

3. Heat losses and heat loss parameter (HLP)
*All plan dimensions are taken to the internal faces of the external walls and
include internal partitions. All heights are taken from finished ground floor level
to the highest ceiling and include any upper floor. Wall areas are net and do not
include the area of any door or window openings.*

Heat losses due to:

External walls	net area (m²) × 0.45	☐ (7)
Ground floor	area (m²) × 0.45	☐ (8)
Roof	area (m²) × 0.25	☐ (9)
Windows	0.9 × area (m²) × 3.30	☐ (10)
Doors	area (m²) × 3.10	☐ (11)
Ventilation	0.33 × (3) × (6)	☐ (12)
Total specific heat loss	(7)+(8)+(9)+(10)+(11)+(12)	☐ (13)
Heat loss parameter	$\dfrac{(13)}{5}$	☐ (14)

4. Water heating energy requirements

Energy content of heated water (Table A1, col. (a)): ⬚ (15)

Distribution loss (Table A1, col. (b)): ⬚ (16)

Required output from boiler: (15)+(16)+2.55 ⬚ (17)

Efficiency of boiler:
Gas with fan-assisted flue	68%
Gas with balanced or open flue	65%
Oil fired	70%

⬚ (18)

Fuel energy required for water heating: $\dfrac{(17)}{(18)} \times 100$ ⬚ (19)

Table A1   Hot water energy requirements and heat gains

Total floor area	Hot water energy requirements		(c)
	(a)	(b)	Heat gains due to lighting, cooking etc.
	Hot water usage	Distribution loss	
m²	GJ/year	GJ/year	W
60	5.68	1.00	382
70	6.17	1.09	431
80	6.65	1.17	480
90	7.11	1.26	528
100	7.57	1.34	576
110	8.01	1.41	623
120	8.44	1.49	669
130	8.86	1.56	715
140	9.26	1.63	760
150	9.65	1.70	805
160	10.03	1.77	849
170	10.40	1.84	893
180	10.75	1.90	935
190	11.10	1.96	978
200	11.43	2.02	1020

(*Source:* Table 1 and Table 5 of Appendix G to Approved Document L: 1995 Edition)

5. Heat gains

Heat gains from water heating: $0.8 \times [(16) + 2.55] + \dfrac{(15)}{4}$ ⬚ (20)

Heat gains from lights, appliances etc. (Table A1 col. (c)): [ ] (21)

Solar gains: window area × 15 [ ] (22)

Total heat gains: (20) + (21) + (22) [ ] (23)

Gains/loss ratio: $\dfrac{(23)}{(13)}$ [ ] (24)

Utilisation factor (Table A2): [ ] (25)

Table A2 Utilisation factor as a function of the gain/loss ratio (G/L)

G/L	Utilisation factor	G/L	Utilisation factor
1	1.00	16	0.68
2	1.00	17	0.65
3	1.00	18	0.63
4	0.99	19	0.61
5	0.97	20	0.59
6	0.95	21	0.58
7	0.92	22	0.56
8	0.89	23	0.54
9	0.86	24	0.53
10	0.83	25	0.51
11	0.81	30	0.45
12	0.78	35	0.40
13	0.75	40	0.36
14	0.72	45	0.33
15	0.70	50	0.30

(*Source:* Table 7 of Appendix G to Approved Document L: 1995 Edition)

Useful gains: (23) × (25) [ ] (26)

6. Mean internal temperature

Mean temperature of living area (Table A3, col. (a)): [ ] (27)

Adjustment for gains: $\left[\dfrac{(26)}{(13)} - 4.0\right]$ [ ] (28)

Adjusted living room temperature: (27) + (28) ⬜ (29)

Temperature difference between zones
(Table A3, col. (b)): ⬜ (30)

Sleeping area fraction: $\dfrac{\text{Total floor area (2)}}{\text{area of zone 2}}$ ⬜ (31)

Mean internal temperature: (29) − [(30) × (31)] ⬜ (32)

Table A3 Living area temperature and the difference between zones

Heat loss parameter (Item 14 on the worksheet)	(a) Mean internal temperature of living area	(b) Difference in temperature between zones
1.0 (or lower)	18.88	1.41
1.5	18.88	1.49
2.0	18.85	1.57
2.5	18.81	1.65
3.0	18.74	1.72
3.5	18.62	1.79
4.0	18.48	1.85
4.5	18.33	1.90
5.0	18.16	1.94
5.5	17.98	1.97
6.0 (or higher)	17.78	2.00

(*Source:* Tables 8 and 9 of Appendix G to Approved Document L: 1995 Edition)

7. Degree-days

Temperature rise from gains: $\dfrac{(26)}{(13)}$ ⬜ (33)

Base temperature: (32) − (33) ⬜ (34)

Degree days: (Table A4) ⬜ (35)

8. Space heating requirements

Energy requirement: 0.0000864 × (35) × (13) ⬜ (36)

Fuel energy for space heating: 0.9 × (36) × $\dfrac{100}{(18)}$ ⬜ (37)

Table A4   Degree days as a function of base temperature

Basse temperature deg C	Degree-days	Base temperature deg C	Degree-days
1.0	0	11.0	1140
1.5	30	11.5	1240
2.0	60	12.0	1345
2.5	95	12.5	1450
3.0	125	13.0	1560
3.5	150	13.5	1670
4.0	185	14.0	1780
4.5	220	14.5	1900
5.0	265	15.0	2015
5.5	310	15.5	2130
6.0	360	16.0	2250
6.5	420	16.5	2370
7.0	480	17.0	2490
7.5	550	17.5	2610
8.0	620	18.0	2730
8.5	695	18.5	2850
9.0	775	19.0	2970
9.5	860	19.5	3090
10.0	950	20.0	3210
10.5	1045	20.5	3330

(*Source:* Table 10 of Appendix G to Approved Document L: 1995 Edition)

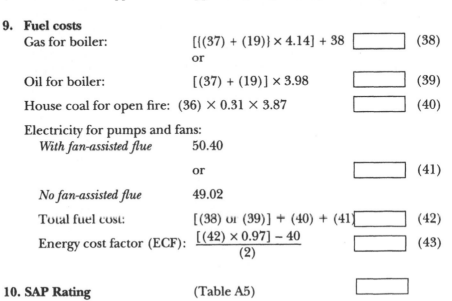

9. **Fuel costs**

Gas for boiler: $[\{(37) + (19)\} \times 4.14] + 38$ ☐ (38)

or

Oil for boiler: $[(37) + (19)] \times 3.98$ ☐ (39)

House coal for open fire: $(36) \times 0.31 \times 3.87$ ☐ (40)

Electricity for pumps and fans:
  With fan-assisted flue   50.40

    or   ☐ (41)

  No fan-assisted flue   49.02

Total fuel cost: $[(38) \text{ or } (39)] + (40) + (41)$ ☐ (42)

Energy cost factor (ECF): $\dfrac{[(42) \times 0.97] - 40}{(2)}$ ☐ (43)

10. **SAP Rating**   (Table A5)   ☐

Table A5   SAP Rating by energy cost factor

Energy cost factor £/m²	SAP Rating
1.4	100
1.5	97
1.6	95
1.7	92
1.8	89
1.9	87
2.0	85
2.2	81
2.4	77
2.6	74
2.8	70
3.0	67
3.3	63
3.6	59

(*Source:* Table 14 of Appendix G to Approved Document L: 1995 Edition)

WORKED EXAMPLE OF SAP ENERGY RATING CALCULATION WORKSHEET

This worksheet is an abbreviated form of the worksheet given in Appendix G.

The property used as the basis for this calculation is the plan shown in Fig. 1.2 except that the garage is taken to be a separate building.

Specification:

Of the property covered by this calculation example:

Two storeys high.

Semi-detached with separate garage building.

One bathroom and separate W.C.

Total floor area 82.50 m².

External walls of bricks and blocks, insulated to a U value of 0.45 W/m²K.

Concrete ground floor insulated to a U value of 0.45 W/m²K.

Timber framed pitched and tiled roof insulated to 0.25 W/m²K.

Timber windows with normal 6 mm double glazing to a U value of 3.3 W/m²K.

Timber external doors half glazed with 6 mm double glazing to a U value of 3.1 W/m²K.

Draught lobbies provided to external doors.

Natural ventilation (except extract fans in bathrooms etc.).

Hot water radiators with thermostatic valves, heated by a gas boiler with fan-assisted flue, controlled by a programmer and a room stat.

Standard 120 litre cylinder with 35 mm factory applied foam lagging and cylinder stat.

All primary pipework insulated.

Two heating zones: living areas (zone 1) and sleeping areas (zone 2).

Solid fuel open fire in lounge.

Gas price £4.14/GJ plus £38 standing charge, coal £3.87/GJ.

Calculation of SAP Rating:

1. Sizes of the dwelling

All plan dimensions are taken to the internal faces of the external walls and include internal partitions. All heights are taken from finished ground floor level to the highest ceiling and include any upper floor.

Ground floor area: $\boxed{41.25}$ (1)

Total floor area: $\boxed{82.50}$ (2)

Total volume: $\boxed{206.25}$ (3)

2. Ventilation rate

Air changes per hour: $\dfrac{80}{(3)}$ $\boxed{0.39}$ (4)

Additional structural and window infiltration:

0.45 *for two storey or* 0.35 *for single storey* $\boxed{0.45}$ (5)

Air change rate: $[(4) + (5)] \times 1.85$ $\boxed{1.55}$ (6)

3. Heat losses and heat loss parameter (HLP)

All plan dimensions are taken to the internal faces of the external walls and include internal partitions. All heights are taken from finished ground floor level to the highest ceiling and include any upper floor. Wall areas are net and do not include the area of any door or window openings.

Heat losses due to:

External walls   net area (79.32 m²) $\times$ 0.45 $\boxed{35.69}$ (7)

Ground floor   area (41.25 m²) $\times$ 0.45 $\boxed{18.56}$ (8)

Roof   area (41.25 m²) $\times$ 0.25 $\boxed{10.31}$ (9)

Windows   0.9 $\times$ area (14.70 m²) $\times$ 3.30 $\boxed{43.66}$ (10)

Doors   area (3.78 m²) $\times$ 3.10 $\boxed{11.72}$ (11)

Ventilation   0.33 $\times$ (3) $\times$ (6) $\boxed{105.50}$ (12)

Total specific heat loss   (7)+(8)+(9)+(10)+(11)+(12) $\boxed{225.44}$ (13)

Heat loss parameter $\dfrac{(13)}{5}$ $\boxed{45.09}$ (14)

4. Water heating energy requirements

Energy content of heated water (Table A1, col. (a)): $\boxed{6.65}$ (15)

Distribution loss (Table A1, col. (b)): $\boxed{1.17}$ (16)

Required output from boiler: (15)+(16)+2.55 $\boxed{10.37}$ (17)

Efficiency of boiler:
 Gas with fan-assisted flue   68%
 Gas with balanced or open flue   65% $\boxed{68\%}$ (18)
 Oil fired   70%

Fuel energy required for water heating: $\frac{(17)}{(18)} \times 100$ $\boxed{15.25}$ (19)

5. **Heat gains**
Heat gains from water heating: $0.8 \times [(16) + 2.55] + \frac{(15)}{4}$ $\boxed{4.64}$ (20)

Heat gains from lights, appliances etc. (Table A1 col. (c)): $\boxed{480}$ (21)

Solar gains: window area $\times$ 15 $\boxed{220.50}$ (22)

Total heat gains: (20) + (21) + (22) $\boxed{705.14}$ (23)

Gains/loss ratio: $\frac{(23)}{(13)}$ $\boxed{3.06}$ (24)

Utilisation factor (Table A2): $\boxed{1.00}$ (25)

Useful gains: (23) $\times$ (25) $\boxed{705.14}$ (26)

6. **Mean internal temperature**
Mean temperature of living area (Table A3, col. (a)): $\boxed{17.78}$ (27)

Adjustment for gains: $\left[\frac{(26)}{(13)} - 4.0\right]$ $\boxed{-0.87}$ (28)

Adjusted living room temperature: (27) + (28) $\boxed{16.97}$ (29)

Temperature difference between zones
(Table A3, col. (b)): $\boxed{2.00}$ (30)

Sleeping area fraction: $\dfrac{\text{Total floor area (2)}}{\text{area of zone 2}}$ $\boxed{2}$ (31)

Mean internal temperature: (29) $-$ [(30) $\times$ (31)] $\boxed{12.85}$ (32)

7. **Degree-days**
Temperature rise from gains: $\frac{(26)}{(13)}$ $\boxed{3.13}$ (33)

Base temperature: (32) $-$ (33) $\boxed{9.72}$ (34)

Degree days: (Table A4) *(Interpolated)* $\boxed{9000}$ (35)

8. **Space heating requirements**
Energy requirement: $0.0000864 \times$ (35) $\times$ (13) $\boxed{17.53}$ (36)

Fuel energy for space heating: $0.9 \times (36) \times \dfrac{100}{(18)}$ $\boxed{\textbf{23.20}}$ (37)

9. Fuel costs

Gas for boiler: $[\{(37) + (19)\} \times 4.14] + 38$ $\boxed{\textbf{197.18}}$ (38)

or

Oil for boiler: $[(37) + (19)] \times 3.98$ $\boxed{}$ (39)

House coal for open fire: $(36) \times 0.31 \times 3.87$ $\boxed{\textbf{21.03}}$ (40)

Electricity for pumps and fans:
With fan-assisted flue 50.40

or $\boxed{\textbf{10.50}}$ (41)

No fan-assisted flue 49.02

Total fuel cost: $[(38) \text{ or } (39) + (40) + (41)$ $\boxed{\textbf{228.71}}$ (42)

Energy cost factor (ECF): $\dfrac{[(42) \times 0.97] - 40}{(2)}$ $\boxed{\textbf{2.20}}$ (43)

10. SAP Rating

(Table 5) (Interpolated) $\boxed{\textbf{81.0}}$

Appendix C

PROVISIONS FOR DISABLED PEOPLE

| 16.1   The Building Regulations applicable |

M2   Access and use of the building

This Regulation has been extended to cover the need to provide certain facilities for the benefit of disabled people in a dwelling, previously it only applied to commercial and public premises.

Where relevant, certain parts of the Regulation, such as special provisions for doors, or the fitting of a WC in the entrance storey, have been dealt with in Appendix D but there are several new requirements that stand alone and are covered in this Appendix.

| 16.2   Access to the building |

The purpose of the legislation is to make it reasonably easy for a disabled person to get from a vehicle, either inside or outside the plot, to an entrance door, preferably the principal entrance. To allow wheelchair users to use this access it should be reasonably level but it is recognized that this is not always possible if the topography is a steep slope. In this case, a stepped approach will be needed but this should be designed with the needs of people using sticks or crutches in mind. A level or ramped driveway may be used as the means of access but it must conform to the standards set out below and be wide enough to provide a path past any parked cars.

| 16.3   The approach path |

The design of the approach paths on a site is determined by what is referred to in Approved Document M2 as 'plot gradient'. This term has nothing to do with the slope of the ground surface but is the gradient from the finished floor level (presumably at the entrance door, although this is not stated) to the point where the access starts, which is either the point of entry to the site or the point where a disabled person would alight from a car within the site.

This means that, where the natural gradient of the ground is critical, the plot gradient can be adjusted by the careful location of the start of the access, the position of the dwelling and the level of the entrance floor. The Approved Document is at pains to point out that the location of the building on the site is a matter for planning legislation rather than a subject within the terms of the Building Regulations and account needs to be taken of compliance in this respect. Planning control would also affect the layout of the car parking provision and highway control may be involved with the point of vehicular entry to the site, all of which may determine whether an access that satisfies the Building Regulations can be achieved.

It is also interesting to note that the *piano nobile* of times long past when the principal entrance floor of the house was elevated well above the street level and was approached via a stately flight of steps would not meet the requirements of this particular Regulation.

Three possibilities are defined in the Approved Document, a level approach, a ramped approach and, where there is no other way to gain access, a stepped approach. In all cases, any cross fall provided on the approach should not exceed 1 in 40.

16.3.1   A level approach

If the path has a gradient of less than 1 in 20, it is considered 'level' and, as such, will satisfy the requirements, provided that the width is at least 900 mm and its surface is firm and smooth enough for the support and manoeuvring of a wheelchair. Loose laid materials, such as gravel or shingle should be avoided since they make access difficult for anybody using sticks or crutches.

16.3.2   A ramped approach

If the plot gradient is greater than 1 in 20 but does not exceed 1 in 15, a ramped path may be provided and would be acceptable if, as in the case of the level approach, the surface is firm and even and the width is at least 900 mm.

The path must be divided into flights by level landings, one at the bottom, one at the top and at intermediate points as required. All landings must be at least 1200 mm long, clear of any gates or doors opening across them.

Between the landings, the flights may have gradients of up to 1 in 15 provided that they are less than 10 m long or 1 in 12 if they are less than 5 m long.

Because of the need to provide landings, the slope of the path must always be steeper than the plot gradient, for instance, the combination of a flight of 1 in 15 extending for 10 m and a level landing of 1200 mm gives a maximum plot gradient of 1 in 18.5, as shown at A in Fig. C.1. The maximum plot gradient of 1 in 15 is exactly the limit produced by a combination of a 5 m ramp at 1 in 12 and a landing of 1200 mm, as shown at B in Fig. C.1.

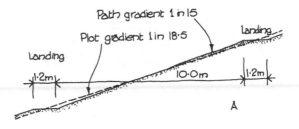

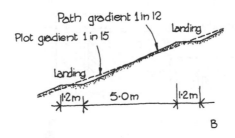

NOTE: Sketches have a
vertical exaggeration of 5:1

Fig. C.1 Access path ramps

If the actual slope of the ground surface follows the plot gradient, it will finish up too high at the building since the latter is set from the finished floor level. In these circumstances it will be necessary to reduce the ground level adjacent to the building for a distance of 2.83 m out from the wall, to provide the 150 mm clearance of the DPC.

16.3.3   A stepped approach

On very sloping sites, a path of less than 1 in 15 may not be achievable, in which case, it is accepted that only the needs of the ambulant disabled have to be accommodated and for this purpose a 'stepped approach' can be used. To satisfy the requirements, a stepped approach should conform to the following:

- minimum width: 900 mm;
- maximum rise of any flight: 1800 mm;
- uniform risers of between 75 mm and 150 mm;
- minimum goings of 280 mm, measured at a point 270 mm in from the narrow end, if the steps are tapered;

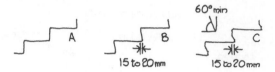

Rise: 75 to 150 mm   Going: 280mm min

SUITABLE STEP PROFILES AS SHOWN IN APPROVED DOCUMENT M2

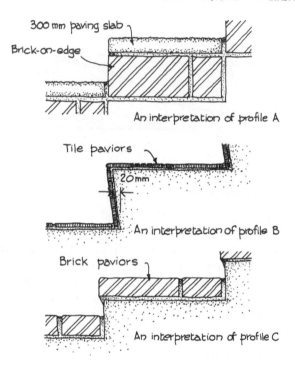

Fig. C.2 Steps in an external approach

- a step profile as shown in Fig. C.2;
- top, bottom and any intermediate landings that are not less than 900 mm long;
- a continuous handrail of a grippable section, between 850 mm and 1000 mm above the pitch line, extending 300 mm beyond the top and bottom nosings.

Having reached the building, it is necessary for the disabled person to be able to get in and, to make this possible, an entrance door of an appropriate width with a level threshold is required. The acceptable requirements for this are set out in the amendments to Chapter 8 of the main text.

16.4 Access within the building

One objective of the legislation is to ensure that, once a disabled person has reached the entrance storey, they can also get into the habitable rooms and to the WC within that storey. No vertical circulation is anticipated except where, in exceptional circumstances, there has to be a change of level within the entrance storey due to severe site conditions. Where there is such a change of level, the needs of an ambulatory disabled person should be met, as described in the amendments of Chapter 9 of the main text.

The requirements can be achieved by the provision of corridors and doorways of adequate width, particularly where a door is approached from the side, and the careful control of local obstructions such as radiators or other fixtures.

The widths of doors and corridors are related because the narrower the door, the wider must the corridor be to enable a wheelchair to be turned sufficiently to get through the opening. The minimum clear width of door is 750 mm; i.e. the distance from the open face of the door to the opposite door stop. To achieve this minimum, all doors to habitable rooms and the entrance level WC must have a leaf width of 830 mm (2 feet 9 inches). The relationship between door width and corridor width is shown in Fig. C.3.

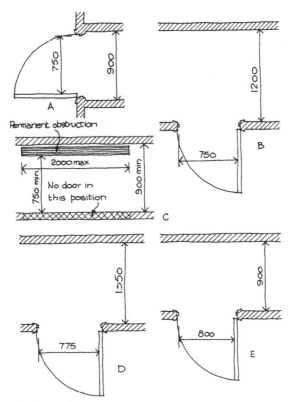

Fig. C.3 Door and corridor widths

Any obstructions that have the effect of reducing the corridor width must not be more than 2 m long and must leave a clear width of at least 750 mm. They should not be located opposite a door, where they would prevent a wheelchair user manoeuvring into the doorway.

16.5   Accessible switches and sockets

To assist those people whose reach is limited by their disability, switches and all types of socket outlets (power, TV, telephone, etc.) should be located at a convenient height.

Approved Document M2 defines a convenient height as between 450 mm and 1200 mm from the floor. It is intended that within this zone, the switch or socket is placed at a suitable height. This means that switches should be fixed slightly lower than past practice, at approximately 1160 mm to the centre line and all sockets raised to approximately 500 mm from the floor to their centre line.

Appendix D

REVISIONS TO THE REGULATIONS AND APPROVED DOCUMENTS

Chapter 1: Introductions

D1.1, page 3

The manual has now been re-issued and gives useful guidance on the Building Regulation system. It explains which building proposals would be subject to the Regulations and the two ways in which they would be controlled. It then goes on to give a review, with explanatory notes of how to meet the requirements set out, including a section on exempt buildings and work.

D1.2, page 3

Since the 1995 Edition of the Regulations, a further revision has been introduced by which Part M has been extended to cover access and facilities for disabled people using houses and flats.

D1.3, page 4

In connection with Regulation Part M, M1 defines the term 'disabled people' as those suffering from an impairment that limits their ability to walk, hear or see or that makes it necessary for them to rely on a wheelchair, crutches, frames or sticks for mobility. It also includes people with arthritis, rheumatism, partial paralysis resulting from a stroke or accident, anybody who finds it difficult to bend their limbs or who has back problems. The definition is such that it encompasses many who would not, otherwise, consider themselves as disabled, merely elderly.

Chapter 8: Windows doors and ventilation openings

D8.1, page 122

There is a further Building Regulation that is now applicable to doors and that is:

M   Access and facilities for disabled people

Regulation M originally applied only to commercial premises but it has now been extended to make access into and within the building safer.

D8.2, page 125

The new Regulation M requires that the minimum width for one of the entrance doors to the house or block of flats should be such as will provide adequate space to allow a wheelchair user to manoeuvre into the dwelling. For this purpose the clear width between the open door face and the opposite door stop should be not less than 775 mm. A standard 830 mm (2 feet 9 inches) door will achieve this.

Where the approach to the house or block of flats is either level or ramped (see 9.8 in the main text), so that a disabled person has been able to reach the entrance door in a wheelchair, it is also necessary to provide an accessible threshold over which the wheelchair wheels can travel. Wherever possible, the approach to the building should either be level or ramped but, should the topography make this impossible and the approach has to be stepped, it would still be reasonable to provide an accessible threshold. If a step at the door cannot be avoided the rise should not be more than 150 mm.

D8.3, page 126

Doorways have always been constructed with steps up to exclude rain and a cill that is above the floor level to allow the door to swing over a door mat. This arrangement is impossible to negotiate in a wheel chair and, therefore, the Regulation M has introduced the concept of accessible thresholds.

Accessible thresholds
Regulation M requires that the threshold of the principal entrance to a house or flat should allow access for wheelchair users and the ambulant disabled. In addition it must, of course, minimise the risk of ingress of any rainwater or damp. Although not mentioned in Approved Document M, the Housing Association has found that the best defence against rain is the provision of a porch or projecting canopy. This could well be considered in severely exposed locations.

The details of the construction of the threshold must, furthermore, comply with all other relevant Regulations such as adequate sub-floor ventilation, soil gas membranes and the prevention of thermal bridging.

Any square upstands, such as a water bar, or a sharp slope impede access and are to be avoided. The maximum upstand or any difference in level between the external landing and the cill, or between the cill and the floor should not exceed 15 mm and should be rounded or chamfered if they are more than 5 mm high. No slopes should be steeper than 15°.

Guidance on the detailing of thresholds is not given in Approved Document M but is to be found in the Stationery Office publication *Accessible thresholds in new housing – Guidance for builders and designers.* In this publication three elements are detailed:

- the external landing;
- the cill and threshold;
- the junction between the threshold and the finished floor.

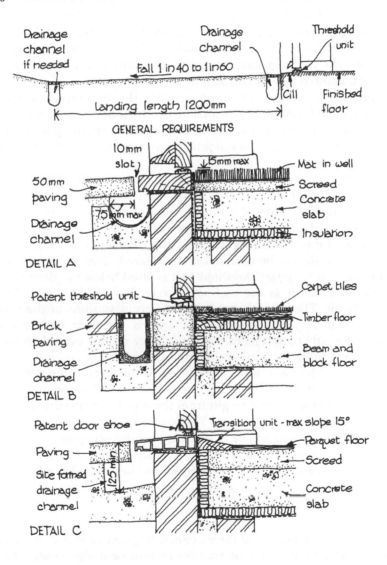

Fig. D.1 Details of accessible thresholds

The landing, as shown in Fig. D.1, should be at least 1200 mm long and 900 mm wide. If the approach to the landing is from one side, the width should be increased to 1000 mm between the face of the wall of the building and any boundary or other obstruction. The inner end of the landing should be level with, or no more than 10 mm below, the front edge of the door cill and fitted with either a site formed drainage slot or a proprietary drainage channel. To reduce the amount of water to be dealt with at the doorway, there should be a fall on the landing of between 1 in 40 and 1 in 60 away from the building.

Two forms of site formed drainage slot are shown in Fig. D.1 at A and C, both using an oversailing paving slab. The ends of the channel should be left open to allow the water to flow out and for rodding purposes. This is likely to occur naturally as the ground or paving level adjacent to door landing will need to be lowered by at least 150 mm to clear the DPC.

If the access path slopes down towards the entrance, an additional drainage channel should be provided across the outer edge of the landing.

The Guide considers that the drainage channels do not need to be connected to a surface water drainage system, unless extremely adverse conditions apply.

Door cills can be made from a variety of materials, Fig. D.1 shows three, timber, stone and plastic. There is a possibility that the raised landing level could cause deterioration of a timber cill, but the Guide considers that a drainage channel constructed as shown, would provide sufficient ventilation to overcome the problem. The slope on the cill should not exceed 15° and the outer edge should be rounded or chamfered. The Guide recommends that a suitable proprietary threshold unit is fitted below the door to prevent any rain penetration as the traditional water bar would make wheelchair access difficult. There are a number of different versions of the threshold unit, as yet there are no formal test criteria for their weather-proofing qualities, the only requirement being that they do not exceed 15 mm high and outer and inner faces are rounded or chamfered.

Careful consideration needs to be given to the level of the finished floor surface in relation to the cill, taking into account any flooring materials to be used so that, if possible, the height from the top of the threshold unit to the flooring does not exceed 15 mm. This could present a problem when the choice of floor covering will be in the hands of the occupant.

Where it is not possible to achieve this ideal, a transition unit can be fitted as shown at C in Fig. D.1, this, however, should be avoided if at all possible.

D8.4, page 126

An additional requirement of the extended Regulation M and its supporting Approved Document is that internal doors must be designed with the needs of disabled people in mind. The following section deals with this.

Internal doors

To facilitate the movement by wheelchair borne people once they are in the dwelling, all doors to habitable rooms on the entrance level and the WC required on this floor must have a clear opening width of at least 750 mm. To achieve this, a standard door leaf of 830 mm (2 feed 9 inches) is required.

This minimum clear width is, however, related to the width of the corridor because of the need to turn the wheelchair into it and may increase to 800 mm, requiring a leaf width of 880 mm. The relationship between corridor width and door width is shown in Fig. C.1 in the new Appendix C.

D8.5, page 127

Approved Document F gives guidance on the way that background ventilation can be achieved in a house or flat. The Document states that it can be provided by means of air bricks, trickle ventilators, vertically sliding sash windows that can be locked open or high level top widows that can also be locked in an open position.

Trickle ventilators should be of a suitable size, fixed either above the window frame or in the head of the frame or be a purpose-made vent fitted between the top of the glazing and the head of the frame.

Air bricks should be of the 'hit-and-miss' type and either have slots of not less than 5 mm width or holes measuring at least 8 mm. These dimensional standards apply to the main air passages, not to insect screens or baffles.

Sash windows can be fitted with pegs that are locked into the sash stiles and limit the extent to which the top sash can be opened. This provision would not be acceptable if the window also had to provide a means of escape.

High-level top-hung windows, often referred to as 'night vents', employed as background ventilation, should be fitted with a lockable fastener with a removable key that is capable of securing the window in at least two positions, the smallest of which providing the required free area. Approved Document F1 recommends that this alternative is only used in windows above the ground floor storey because of the security risk.

Chapter 9: Staircases

D9.1, page 138

The requirements of Regulation M2 now do apply to houses and so are obligatory rather than a sensible standard to follow in domestic work.

D9.2, page 139

The rules set out for a domestic stair apply to a stair in a house or within a flat.

D9.3, page 140

There is no recommendation as to a minimum width for a stair in Part K but Part M states that if the site conditions are such that a stairway is required within the entrance storey of a dwelling the minimum width should be 900 mm to enable an ambulatory disabled person to negotiate the change of level with assistance.

D9.4, page 140

Where there are steps or a stair within the entrance storey of the dwelling there should be a handrail to both sides of a flight of more than three risers and to both sides of any landings for the benefit of ambulatory disabled people.

D9.5, page 140

Additional requirements are imposed by Regulation M with regard to stairs in flats.

Common stairs

In a staircase serving more than one flat and where there is no lift installed, the stairs follow the rules set out in this Chapter except for:

- maximum rise 170 mm;
- minimum going 250 mm (270 mm on a tapered tread);
- the risers must not be of the open type;
- all step nosing must be distinguished by a strip of a contrasting brightness;
- the profile of the steps must follow one of the types shown in Fig. D.2;
- handrails are to be provided to both sides of the stair and continued horizontally at the top and the bottom of the flight for a distance of 300 mm from the line of the nosings at a height of 1000 mm above the landing;
- minimum width – no minimum width is given unless the stair is an escape route, in which case it should be not less than 1 m (see 15.7.7 in the main text).

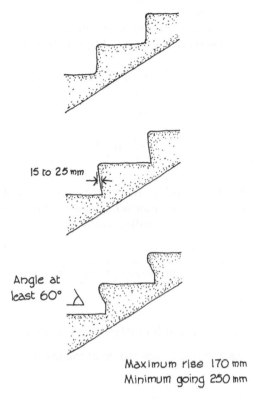

15 to 25 mm

Angle at least 60°

Maximum rise 170 mm
Minimum going 250 mm

Fig. D.2 Profile of steps in a common stairway in a block of flats

D9.6, page 144

In many respects the standards for ramps are the same as for stairs unless the ramp is in the approach to the building in which case the dimensions that apply are set out in the new Appendix C.

D9.7, page 145

Since lifts are the only practical means for disabled people to gain access to an upper storey, the Approved Document M, in support of the extended Regulation M, defines the standards to be followed. These are set out in the following section.

Lifts
The Regulations require that reasonable provision shall be made to allow disable people to use the building and one way of satisfying this, in a block of flats, would be to install a passenger lift.

If such a lift is installed, there should be a clear landing provided, 1500 mm square, in front of the doors.

The lift car should be at least 900 mm wide and 1250 mm long and both the landing doors and the car doors should provide a clear opening width of at least 800 mm.

The door should have a 'dwell time' of five seconds before they begin to close but this can be reduced to not less than three seconds by the use of a suitable electronic door re-activating device, other than a door edge pressure system.

The lift controls, both on the landing and in the car should be positioned between 900 mm and 1200 mm above the floor and be accompanied by a suitable tactile storey indication. There should also be a visual indication on the landing that the lift is coming. In the car there should also be a further tactile indication to confirm the floor selected and, if the lift serves more than three floors, a visual and audible indication of the floor reached.

Chapter 12: Bathrooms and above ground drainage

D12.1, page 171

There is now an additional Building Regulation applicable to this work:

M3(1) WC provisions in the entrance storey of the dwelling

D12.2, page 171

Regulation M3(1) introduces the requirement to provide toilet facilities for disabled people at the entrance level of the dwelling.

D12.3, page 171

For the benefit of disabled people there should be a toilet located in the entrance storey.

D12.4, page 173

The toilet at the entrance level of the dwelling, provided for the convenience of disabled people, should have the door opening outwards with a clear opening width of at least 750 mm (more is preferable) and be positioned to enable wheelchair users to gain access to the WC. Approved Document M recognizes that it will not always be possible to fit the wheelchair fully into the compartment.

Adequate space must be provided to allow access to the WC and the washbasin and the latter should not obstruct the access to the former.

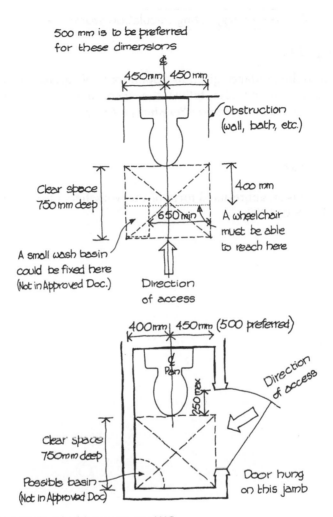

500 mm is to be preferred
for these dimensions

450mm | 450mm

Obstruction
(wall, bath, etc.)

Clear space
750 mm deep

400 mm

650 min

A wheelchair
must be able
to reach here

A small wash basin
could be fixed here
(Not in Approved Doc.)

Direction
of access

400mm | 450mm (500 preferred)

₵
Pan

250 max

Direction
of access

Clear space
750mm deep

Possible basin
(Not in Approved Doc)

Door hung
on this jamb

Fig. D.3 Clear space for wheelchair access to a W.C.

Fig. D.3 shows the desirable clear space for a wheelchair when the approach to the WC is from the front. If there is an oblique line of access, the dimension from the centre line to the side of the clear space opposite to the direction of approach can be reduced to 400 mm if necessary.

Chapter 15: Means of escape and fire fighting facilities

D15.1, page 233

In addition to the guidance given in Approved Document K, the guarding of an escape route over a flats roof is covered in Approved Document M.

Appendix A   SAP energy rating calculation worksheet

DA. I, page 240

In addition to the guidance given in Appendix G of Approved Document L, the SAP rating should also follow the recommendations given in the Government's *Standard Assessment Procedure for Energy Rating of Dwellings: 1998 Edition*.

DA.2, page 240

The worksheet is an abbreviated form of the worksheet given in Appendix G as adjusted by SAP 1998.

INDEX